Fundamentals of Digital Electronics

Fundamentals of Digital Electronics

Bryanna Bolt

NY RESEARCH PRESS
New York

Published by NY Research Press
118-35 Queens Blvd., Suite 400,
Forest Hills, NY 11375, USA
www.nyresearchpress.com

Fundamentals of Digital Electronics
Bryanna Bolt

International Standard Book Number: 978-1-64725-425-4 (Hardback)

Cataloging-in-Publication Data

Fundamentals of digital electronics / Bryanna Bolt.
p. cm.
Includes bibliographical references and index.
ISBN 978-1-64725-425-4
1. Digital electronics. 2. Electronic systems. 3. Electronics. I. Bolt, Bryanna.
TK7868.D5 D54 2023
621.381--dc23

Contents

Preface

This book has been a concerted effort by a group of academicians, researchers and scientists, who have contributed their research works for the realization of the book. This book has materialized in the wake of emerging advancements and innovations in this field. Therefore, the need of the hour was to compile all the required researches and disseminate the knowledge to a broad spectrum of people comprising of students, researchers and specialists of the field.

Digital electronics refers to a branch of electronics concerned with the study of digital signals as well as components which use or generate them. Digital electronic circuits are typically constructed from large assemblies of logic gates which are packed in integrated circuits. The digital circuit is made up of several components serving a specific purpose. These components are classified into two types, including active components and passive components. The active components include diodes and transistors, whereas, the passive components comprise inductors, capacitors and resistors. Digital circuits are employed in a variety of fields and are an essential component of electronic gadgets. There are various applications of digital electronics including quantum computing and rocket science, traffic lights, display of digital watches and operations of automatic doors. This book unravels the recent studies on digital electronics. It will help the readers in keeping pace with the rapid changes in this field.

At the end of the preface, I would like to thank the authors for their brilliant chapters and the publisher for guiding us all-through the making of the book till its final stage. Also, I would like to thank my family for providing the support and encouragement throughout my academic career and research projects.

Bryanna Bolt

1

Number System, Boolean Algebra and Logic Gates

1.1 Number System: Introduction to Various Number Systems and their Conversion

Several number systems are in use in digital technology. The most common are the decimal, binary, octal and hexadecimal systems. The decimal system is typically the most familiar to us because it is a tool that we use every day.

Types of Number Systems

- Binary Number system.
- Decimal Number system.
- Hexadecimal Number system.
- Octal Number system.

Decimal	Binary	Octal	Hexadecimal
0	0000	0	0
1	0001	1	1
2	0010	2	2
3	0011	3	3
4	0100	4	4
5	0101	5	5
6	0110	6	6
7	0111	7	7
8	1000	10	8
9	1001	11	9
10	1010	12	A
11	1011	13	B
12	1100	14	C

13	1101	15	D
14	1110	16	E
15	1111	17	F

Binary Number

- A decimal number such as 7392 can be represented as,

  $7 \times 10^3 + 3 \times 10^2 + 9 \times 10^1 + 2 \times 10^0$

- The number with a decimal point is represented by a series of coefficients as follows,

  $a_5\, a_4\, a_3\, a_2\, a_1\, a_0\, a_1\, a_{-2}\, a_{-3}$

- Decimal equivalent of the binary 11010.11 is 26.75,

  $1 \times 2^4 + 1 \times 2^3 + 0 \times 2^2 + 1 \times 2^1 + 0 \times 2^0 + 1 \times 2^{-1} + 1 \times 2^{-2} = 26.75$

- The number expressed in base-r system has coefficients multiplied by powers of r,

  $a_n \cdot r^n + a_{n-1} \cdot r^{n-1} + \ldots + a_2 \cdot r^2 + a_1 \cdot r + a_0 + a_{-1} \cdot r^{-1} + a_{-2} \cdot r^{-2} + \ldots + a_{-m} \cdot r^{-m}$

Where r = 2, 3, 4... 8, 9, 10 ...16...

System	Radix	Allowable Digits
Binary	2	0,1
Octal	8	0,1,2,3,4,5,6,7
Decimal	10	0,1,2,3,4,5,6,7,8,9
Hexadecimal	16	0,1,2,3,4,5,6,7,8,9,A, B, C, D, E, F

$$
\begin{aligned}
(4021.2)_5 &= 4 \times 5^3 + 0 \times 5^2 + 2 \times 5^1 + 1 \times 5^0 + 2 \times 5^{-1} \\
&= 4 \times 125 + 0 + 10 + 1 + 2 \,x(1/5) \\
&= 500 + 11 + .4 \\
&= (511.4)_{10}
\end{aligned}
$$

$$
\begin{aligned}
(B65F)_{16} &= 11 \times 16^3 + 6 \times 16^2 + 5 \times 16^1 + 15 \times 16^0 \\
&= 11 \times 4096 + 6 \times 256 + 5 \times 16 + 15 \\
&= 45056 + 1536 + 80 + 15 \\
&= (46687)_{10}
\end{aligned}
$$

- Augend: 101101
- Addend: +100111
- 1010100
- Minuend: 101101
- Subtrahend: – 100111
- 000110
- Multiplicand: 1011
- Multiplier: × 101
- 110111

Number Base Conversions

- A binary number can be converted to decimal by forming sum of powers of 2 of those coefficients whose value is 1,

  $(1010.011)_2 = 2^3 + 2^1 + 2^{-2} + 2^{-3} = (10.375)_{10}$

- Likewise, a number expressed in base *r* can be converted to its decimal equivalent by multiplying each coefficient with the corresponding power of *r* and adding,

  $(630.4)_8 = 6 \times 8^2 + 3 \times 8^1 + 0 \times 8^0 + 4 \text{ x } 8^{-1} = (408.5)_{10}$

- Conversion from Decimal 41 to Binary:

Integer	Quotient Remainder	Coefficient
41/2	20+½	$a_0 = 1$
20/2	10+0	$a_1 = 0$
10/2	5+0	$a_2 = 0$
5/2	2+½	$a_3 = 1$
2/2	1+0	$a_4 = 0$
1/2	0+½	$a_5 = 1$

- The conversion from decimal integers to any base-r system is similar to example, except that division is done by r instead of 2.
- Conversion from Decimal 153 to Octal,

153			
19	1		
2		3	
0			2 $=(231)_8$

- Conversion from Decimal fraction $(0.6875)_{10}$ to Binary:

Integer	Fraction	Coefficient
0.6875 x 2	1 + 0.3750	$a_1 = 1$
0.3750 x 2	0 + 0.7500	$a_2 = 0$
0.7500 x 2	1 + 0.5000	$a_3 = 1$
0.5000 x 2	1 + 0.0000	$a_4 = 1$

Multiplication is by r instead of 2 and the coefficients found from integers may range in value from 0 to r-1 instead of 0 and 1.

Conversion from Decimal fraction $(0.513)_{10}$ to Octal,

$0.513 \times 8 = 4.104$

$0.104 \times 8 = 0.832$

$0.832 \times 8 = 6.656$

$0.656 \times 8 = 5.248$

$0.248 \times 8 = 1.984$

$0.984 \times 8 = 7.872$

$(0.513)_{10}=(0.406517...)_8$

The conversion of decimal numbers with both integers and fraction parts is done by converting the integer and the fraction separately and then combining the two answers.

Octal and Hexadecimal Numbers

The conversion from binary, octal and hexadecimal plays an important part in digital computers. Since $2^3 = 8$ and $2^4 = 16$, each octal digit corresponds to three binary digits and each hexadecimal digit corresponds to four binary digits.

- Conversion from binary to Octal,

  $(10\ 110\ 001\ 101\ 011.\ 111\ 100\ 000\ 110)_2 = (26153.7406)_8$

- Conversion from binary to Hexadecimal,

  $(10\ 1100\ 0110\ 1011.\ 1111\ 0000\ 0110)_2 = (2C6B.F06)_{16}$

- Conversion from Octal to binary,

  $(673.124)_8 = (110\ 111\ 011.\ 001\ 010\ 100)_2$

- Conversion from Hexadecimal to binary,

  $(306.D)_{16} = (0011\ 0000\ 0110.\ 1101)_2$

- Conversion from Hexadecimal to Decimal,

  $(37B)_{16}$

$$\begin{aligned} 3\times16^2+7\times16^1+11\times16^0 &= 3\times256+7\times16+11\times1 \\ &= 768+112+11 \\ &= (891)_{10} \end{aligned}$$

Applications of Octal Number System

The applications of octal number system are:

- It is used for entering, the binary data and displaying certain information's.
- It is very important for the efficient use of microprocessors and other digital circuits.

Decimal Number System

The decimal number system uses the base 10 and consists of 10 digits such as 0 to 9. Numbers greater than 9 are described using positional weights. Starting at the Least Significant Digit (LSD), the positional value is 10^0 or 1 with the next positional value being 10^1 or 10. For each next Most Significant Digit (MSD), the positional value goes up by a factor of 10.

For example, the number 1024_{10} can be represented by multiplying each digit in the number by its positional value,

$$1\times10^3+0\times10^2+2\times10^1+4\times10^0 = 1024_{10}$$

Converting Hex to Decimal

To convert to a decimal number use the following steps:

- Find the decimal equivalent for each Hex digit.
- Multiply each decimal value by the Hex positional value.
- Add the results from step 2 to find the decimal value.

As an example, convert $3AF_{16}$ to decimal.

- We can determine that,

  $3^{16} = 3_{10}$ $A^{16} = 10_{10}$ $F^{16} = 15_{10}$

- Multiply each decimal value by the Hex positional value. Add the results to find the decimal equivalent,

  $3 \times 16^2 + 10 \times 16^1 + 15 \times 16^0 = 943_{10}$

Converting Hex to Binary

To convert a Hex number to a binary number use the following steps:

- Write down the Hex number.
- Write below the number, the 4 bit binary equivalent for each Hex digit.
- Combine the 4 bit numbers into one binary number.

As an example, convert $3AF_{16}$ to binary.

Converting Hex to Binary:

3	A	F	Hex Number
0011	1010	1111	4-Bit Equivalent
001110101111			Combined Bits

1.2 Arithmetic Operation using 1's and 2's Compliments

Addition

Boolean addition corresponds to the logical function of an OR/gate as well as to parallel switch contacts,

0 + 0 = 0

0 + 1 = 1

1 + 0 = 1

1 + 1 = 1

The first 3 sums create perfect sense to anyone familiar with elementary addition. The Last sum, though, is quite possibly responsible for more confusion than any other single statement in digital electronics, because it seems to run contrary to the basic principles of mathematics. It does contradict principles of addition for real numbers, but not for Boolean numbers.

In Boolean algebra, there are only two possible values for any quantity and for any arithmetic operation: 1 or 0. There is no such thing as 2 within the scope of Boolean values. Since the sum "1 + 1" certainly is not 0, it should be 1 by process of elimination.

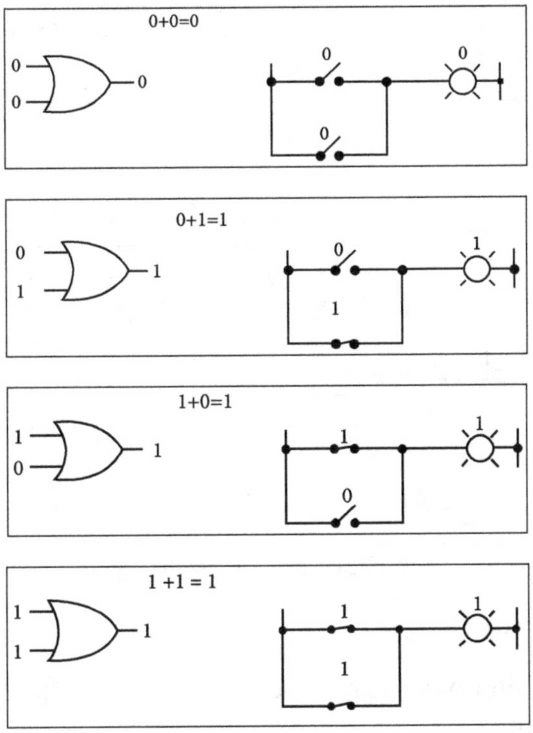

Addition – OR Gate Logic.

Subtraction

Subtraction Implies the existence of negative numbers: 5 - 3 is the same thing as 5 + (−3) and in Boolean algebra, negative quantities are forbidden. There is no such thing as division in Boolean mathematics. Since division is really nothing more than compounded subtraction, in the same way that multiplication is compounded addition.

Multiplication

Multiplication is valid in Boolean algebra and it is the same as in real-number algebra. Anything multiplied by 0 is 0 and anything multiplied by 1 remains unchanged,

$0 \times 0 = 0$

$0 \times 1 = 0$

$1 \times 0 = 0$

$1 \times 1 = 1$

In other words, Boolean multiplication corresponds to the logical function of an AND gate, as well as to series switch contacts.

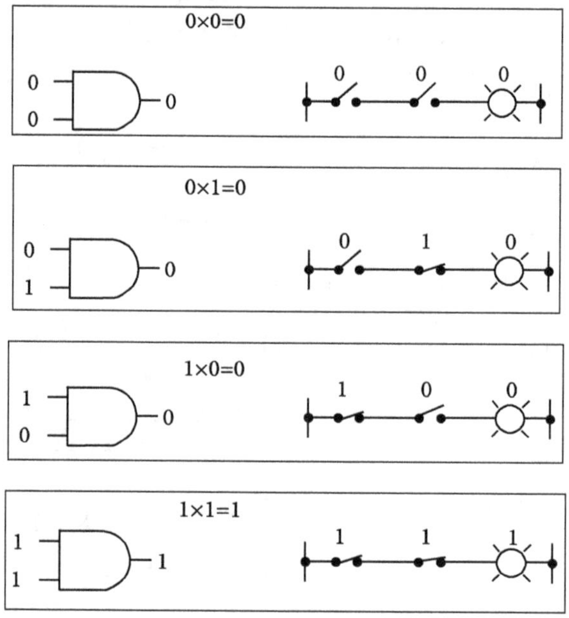

Multiplication – AND Gate logic.

Complementary Function

Boolean complementation finds equivalency in the form of the NOT gate or a normally closed switch or relay contact.

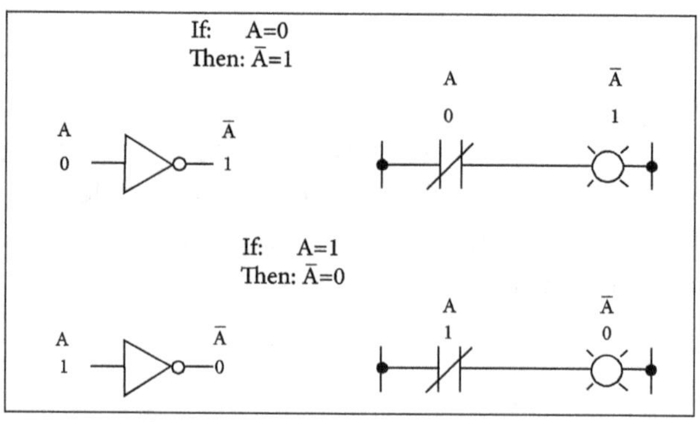

Complementary Function – NOT gate Logic.

1's Complement (Diminished Radix Complement)

All '0's become '1's.

All '1's become '0's.

Example: $(10110000)_2 = (01001111)$

2's Complement (Radix Complement)

Take 1's complement and then add 1.

or,

Toggle all bits to the left of the first '1' from the right.

1.2.1 Signed Binary and Floating Point Number Representation

It is customary to represent the sign with a bit placed in the leftmost position of the number and in order to make it 0 for positive and 1 for negative.

Consider the number 9 is represented in binary with 8 bits. Thus, the +9 is represented with sign bit 0 in the leftmost position which is followed by the binary equivalent of 9 to give 00001001.

Example:

- 00001001.
- Signed magnitude ◇ 10001001.
- Signed-1's complement ◇ 11110110.
- Signed-2's complement ◇ 11110111.

Floating Point Representation

The floating point describes a system for numerical representation in which a string of digits are used to represent the rational number.

The term floating point refers to the fact that the radix point can "float" anywhere in relation to the significant digits of the number. This position can be separately indicated in the internal representation and floating-point representation can be a computer realization. The main advantage of the floating-point representation over fixed-point representation is that it can support a much wider range of values.

For example, a fixed-point representation that has seven decimal digits, with the decimal point assumed to be positioned after fifth digit, can represent the numbers 12435.6, 8523.43, 165.00 and so on, whereas a floating-point representation (such as IEEE 754 decimal 32 format) with seven decimal digits could represent 1.24356, 124356.7, 0.00001243567, and 1243567000000000 and so on. The floating-point format slightly needs more storage to encode the position of radix point, so when stored in the same space, floating-point numbers achieve their greater range at the expense of slightly lesser precision.

For example,

01000000011000000000000000000000

$$=+(1.11)_2 \times 2^{(128-127)}$$

$$=+(1.11)_2 \times 2^1$$

$$=+(11.1)$$

$$=+\left(1\times2^1+1\times2^0+1\times2^{-1}\right)=(3.5)_{10}$$

00111111011000000000000000000000

Floating-Point Numbers

- It have a range that is limited by the significant digits used to represent the number.
- For some applications, it is often necessary to deal with numbers which are very large or very small.
- For these cases, it is better to use a floating-point representation in which numbers are represented by a mantissa comprising the significant digits and an exponent of the radix R.
- The format is Mantissa x $R^{Exponent}$.
- The numbers are usually normalized such that the radix point is placed to the right of first non-zero digit (for example, 5.234×10^{43} or 3.75×10^{-35}).

1.2.2 Introduction to Binary Codes and their Applications

The binary code is the process of assigning the bit patterns to a discrete elements of information. There can be different ways of assigning the bit patterns to this information, one way could be the straight forward assigning of binary codes which are binary equivalent of their numeric value. Another approach could be assigning some other combination of bits.

Table: Two different ways of assigning binary codes:

Decimal code	Binary code format – 1	Binary code format – 2
0	0000	0000
1	0001	0010

2	0010	0100
3	0011	0110
4	0100	1000
5	0101	1001
6	0110	0111
7	0111	0101
8	1000	0011
9	1001	0001

The relation between the number of bits in code word and the number of unique code words is, $N = 2^j$.

Where,

j is the number of bits.

N is the total number of code words.

The binary code can be of following types:

- Weighted Binary Codes: The Weighted Binary Codes are of 8421 or BCD code, 2421 code etc.
- Non-weighted binary code: Non-weighted binary code are gray code.

Binary Coded Decimal Codes

BCD codes are the weighted binary codes. The other examples of weighted codes are 2421, 5421, 5411 code. These codes obey their positional weighting principles. The bits are multiplied by their weights, the sum of these weighted binary bits gives the equivalent decimal digit in the weighted binary system. All the BCD codes are used to represent the decimal numbers. Some other examples of BCD codes are excess-3 Code (XS–3), 84-2-1 code (+ 8, + 4, −2, −1), Biquinary code, 7421 code, 5311 code etc.

Table: Example of BCD codes:

0	**XS-3**	**8421**	**8421**	**Biquinary ABCDEFG**	**2421**	**5311**	**5421**
0	0011	0000	0000	0100001	0000	0000	0000
1	0100	0001	0111	0100010	0001	0001	0001
2	0101	0010	0110	0100100	0010	0011	0010
3	0110	0011	0101	0101000	0011	0100	0011
4	0111	0100	0100	0110000	0100	0101	0100

5	1000	0101	1011	1000001	1000	0111	1000
6	1001	0110	1010	1000010	1001	1001	1001
7	1010	0111	1001	1000100	1010	1011	1010
8	1011	1000	1000	1001000	1011	1100	1011
9	1100	1001	1111	1010000	1100	1101	1100

The BCD codes has certain properties which make it very distinctive and thus have the advantage over the other code. These properties are:

- Self-complementing property.
- Reflective property.

Self-Complementing Codes

In the BCD system, some code have self-complementing property. i.e., their logical complement is the same as its arithmetic equivalent. If the 9's complement of XS-3 is the same as its logical complement helps to find the extensive use in decimal arithmetic.

Examples: 6311 code, 2421 code, XS-3 code.

We can verify the self-complementing property of these codes from the above table.

Reflective Codes

This code is reflected about the center entries with one bit changed. It is said that the 9's complement of a reflected BCD code word is formed simply by changing only one of its bit.

Examples: 2421 code, 5211 code and XS-3 code.

Advantages of BCD Codes

- It is similar to the decimal system.
- The binary equivalent of decimal numbers 0 to 9 only.

Disadvantages of BCD Codes

- The BCD arithmetic is little more complicated.
- The addition and subtraction of BCD have different rules.
- BCD needs more number of bits than the binary for representing the decimal number. So, the BCD is less efficient than binary.

Applications of Binary Number System

The most common application for the binary number system can be found in computer technology. All the computer language and programming is based on the 2-digit number system used in digital encoding. Digital encoding is the process of taking data and representing it with discrete bits of information. These discrete bits consist of the 0s and 1s of the binary system.

For example, the images we see on our computer screen have been encoded with a binary line for each pixel. If a screen is using a 16-bit code, then each pixel has been told what color to display based on which bits are 0s and which bits are 1s. As a result, 2^16 represents 65,536 different colors.

1.3 Boolean Algebra and Identities

Boolean algebra is the mathematics which is used to analyze the digital gates and circuits. These "Laws of Boolean" can be used to reduce both and simplify the complex Boolean expression in an attempt to reduce the number of logic gates required. The Boolean algebra is therefore, a system of mathematics which is based on logic that has its own set of rules or laws that are used to define and reduce Boolean expressions.

The variables which are used in Boolean Algebra only have one of two possible values, a logic "0" and a logic"1", but an expression can have the infinite number of variables all labeled individually to represent the inputs to the expression. For example, the variables A, B, C etc., give us a logical expression of A + B = C, but each variable can ONLY be 0 or 1.

The examples for these individual laws of Boolean, rules and theorems of Boolean algebra are given in the following table.

Truth Tables for the Laws of Boolean:

Boolean Expression	Description	Equivalent Switching Circuit	Boolean Algebra Law or Rule
A + 0 = A	A in parallel with open = "A"	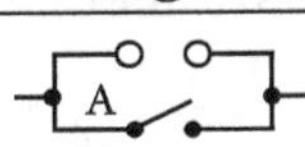	Identity
A + 1 = 1	A in parallel with closed = "CLOSED"	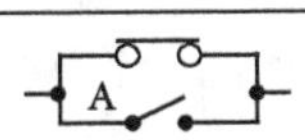	Annulment
A . 0 = 0	A in series with open = "OPEN"	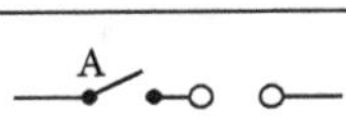	Annulment
A . 1 = A	A in series with closed = "A"	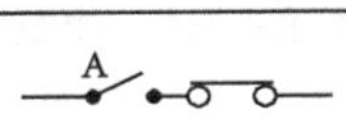	Identity

A . A = A	A in series with A = "A"		Idempotent
NOT $\bar{A}$ = A	NOT NOT A(double negative) = "A"		Double Negation
A + $\bar{A}$ = A	A in parallel with A = "A"	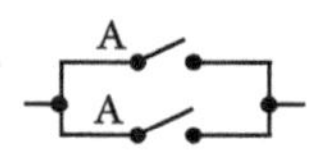	Idempotent
A . $\bar{A}$ = 0	A in series with NOT A = "OPEN"		Complement
A + A = 1	A in parallel with NOT A = "CLOSED"	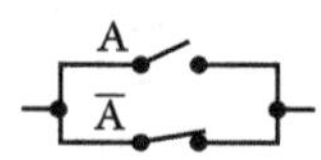	Complement
A + B = B + A	A in parallel with B =B in parallel with A	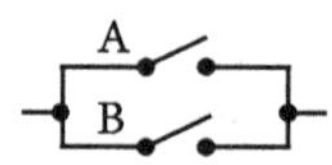	Commutative
$\overline{A+B} = \bar{A}.\bar{B}$	Invert and replace OR with AND		De Morgan's Theorem
A.B = B.A	A in series with B =B in series with A		Commutative
$\overline{A.B} = \bar{A}+\bar{B}$	Invert and replace AND with OR		De Morgan's Theorem

The basic laws of Boolean algebra can relate to the Commutative Law allowing a change in position for the addition and multiplication, the Associative Law allowing the removal of brackets for the addition and multiplication, as well as the Distributive Law allowing the factoring of an expression, are the same as in ordinary algebra.

Every Boolean Laws given above is just a single or two variables, but the number of variables defined is by a single law which is not limited to this, as there can be an infinite number of variables as inputs to the expression. Those Boolean laws that are detailed above can be used to prove any given Boolean expression as well as for simplifying the complicated digital circuits.

A brief description of the various Laws of Boolean are given below representing a variable input.

Description of the Laws of Boolean algebra

Identity Law

A term OR'ed with the value "0" or AND'ed with the value "1" will be always equal to that term.

Where,

A + 0 = A A variable OR'ed with the value 0 is always equal to the variable.

A . 1 = A A variable AND'ed with the value 1 is always equal to the variable.

Annulment Law

The term AND'ed with a "0" equals 0 or OR'ed with a "1" will equal 1.

Where,

A . 0 = 0 A variable AND'ed with the value 0 is always equal to 0.

A + 1 = 1 A variable OR'ed with the value 1 is always equal to 1.

Idempotent Law

An input that is AND'ed or OR'ed with itself is equal to that input.

Where,

A + A = A A variable OR'ed with itself is always equal to the variable.

A . A = A A variable AND'ed with itself is always equal to the variable.

Complement Law

The term AND'ed with its complement equals to "0" and a term OR'ed with its complement equals to "1".

Where,

A . $\bar{A}$ = 0A variable AND'ed with its complement is always equal to 0.

A + $\bar{A}$ = 1A variable OR'ed with its complement is always equal to 1.

Commutative Law

An order of the application of two separate terms is not important.

Where,

A . B = B . A The order in which two variables are AND'ed makes no difference.

A + B = B + A The order in which two variables are OR'ed makes no difference.

Double Negation Law

The term that is inverted twice is equal to the original term.

Where,

$\overline{\overline{A}}$ = A A double complement of a variable is always equal to the variable.

The other algebraic Laws of Boolean not detailed above include:

Distributive Law

It permits the multiplying or factoring out of an expression.

Where,

A (B + C) = A.B + A.C (OR Distributive Law).

A + (B.C) = (A + B).(A + C) (AND Distributive Law).

Absorptive Law

It enables the reduction in the complicated expression to a simpler one by absorbing like terms.

Where,

A + (A.B) = A (OR Absorption Law).

A(A + B) = A (AND Absorption Law).

Associative Law

It allows the removal of brackets from an expression and regrouping of the variables.

Where,

A + (B + C) = (A + B) + C = A + B + C (OR Associate Law).

A(B.C) = (A.B)C = A . B . C (AND Associate Law).

Boolean Algebra Functions

By using the information above, a simple 2-input and OR and NOT gates could be represented by 16 possible functions as shown in the following table:

Function	Description	Expression
1.	NULL	0
2.	IDENTITY	1
3.	Input A	A
4.	Input B	B
5.	NOT A	$\overline{A}$

6.	NOT B	$\overline{B}$
7.	A AND B (AND)	$A . B$
8.	A AND NOT B	$A.\overline{B}$
9.	NOT A AND B	$\bar{A} . B$
10.	NOT A AND NOT B (NAND)	$\overline{A.B}$
11.	A OR B (OR)	$A + B$
12.	A OR NOT B	$A+\overline{B}$
13.	NOT A OR B	$\bar{A} + B$
14.	NOT OR (NOR)	$\overline{A+B}$
15.	Exclusive-OR	$A.\overline{B}+\overline{A}.B$
16.	Exclusive-NOR	$A.B+\overline{A}.\overline{B}$

Example: Using the above laws, we simplify the following expression: (A + B)(A + C)

Q = (A + B).(A + C).

Q = A.A + A.C + A.B + B.C Distributive law.

Q = A + A.C + A.B + B.C Idempotent AND law (A.A = A).

Q = A(1 + C) + A.B + B.C Distributive law.

Q = A.1 + A.B + B. C Identity OR law (1 + C = 1).

Q = A(1 + B) + B.C Distributive law.

Q = A.1 + B.C Identity OR law (1 + B = 1).

Q = A + (B.C) Identity AND law (A.1 = A).

Then the expression (A + B) (A + C) can be simplified to A + (B.C) as in the Distributive law.

Duality Theorem

The Duality Theorem starts with a Boolean relation. We can derive another Boolean relation by:

- Changing OR (operation) i.e., + (Plus) sign to an AND operation.
- Complement any 0 or 1 appearing in the expression.

1.3.1 Complete Logic set

A functionally-complete set of Boolean function consists of a set of Boolean functions from which we can construct all the Boolean functions. {AND_2, NOT}, {NOR_2, NOT}, {$NAND_2$}, {NOR_2} are four functionally-complete sets.

Considering that there are an infinite number of Boolean functions, they can be built from only 3 Boolean functions.

The set of functions {AND_2, OR_2, NOT} are said to be functionally complete.

Definition

A set of Boolean functions is functionally complete, if all other Boolean functions can be constructed from this set and a set of input variables are provided.

Implementing a Boolean Function with Only AND_2 and NOT

It is very easy to get rid of OR. We only have to apply two properties of Boolean functions.

Double Negation: \\x == x. If we negate any expression, it is functionally equivalent to the expression itself.

De Morgan's Law

There are two forms of this law:

Negation of AND

\(xy) == \x + \y

If we negate the conjunction two Boolean expressions, x and y, this changes the AND to OR, and negates the expressions. This can be verified by truth tables.

Negation of OR

\(x + y) == \x\y

If we negate the disjunction of two Boolean expressions, x and y, this changes the OR to AND, and negates the expressions. This can be verified by truth tables.

Any truth table can be written as a sum of products. This has the following general form,

$$Z = P_1 + P_2 + \ldots + P_1$$

Where P_i consists of literals connected by AND_2. In particular, $P_i = l_1 l_2 \ldots\ldots\ldots l_i$ where l_i is a literal.

The expression Z can be written in equivalent as,

$$Z = NOT\left(NOT(P_1 + P_2 + ... + P_K)\right)$$

This is an application of double-negation.

Then, apply DeMorgan's Law on the inner negation,

$$Z = NOT(\backslash P_1 * ... * \backslash P_K)$$

Notice that the ORs changed to ANDs when applying one level of De Morgan's Law, and that each of the product terms now has a negation. We don't apply DeMorgan's any further. In particular, we leave P_i' alone because we don't want to convert the ANDs in the P_i to ORs.

As we can see, by applying double negation and De Morgan's Law once, there are no more OR_2. We should try applying these laws to show that OR_2 and NOT are also functionally complete.

Analyzing AND_2 and OR_2 are Functionally Complete or not

AND_2 and OR_2 are not functionally complete. Hence, {AND_2, OR_2} is not a functionally complete set. Consider any function composed of these two functions. Suppose we set all the input variables values to 1. Then, the output is always 1, regardless of how we put the function together. Similarly, if we set all the input variables values to 0, the output of the function is always 0.

However, we can clearly define the truth table whose output is 1 when all the input variable's values are all 0's, and whose output is 0 when all the input variable's values are all 1's. We cannot implement such a circuit using only {AND_2, OR_2}.

Thus, {AND_2, OR_2} is not functionally complete.

Using One Boolean Function

It is possible that only a single Boolean function is functionally complete. It turns out {AND_2} and {OR_2} are not functionally complete, for the same reasons that {AND_2, OR_2} is not functionally complete. Also, {NOT} is not functionally complete. We can only create two truth tables by composing NOT, and both the functions have a single bit of input.

We know that if there is a single function, it must have at least two inputs, otherwise, we could not construct the Boolean functions with two or more inputs. There are two functionally complete sets that have only one function in it such as {$NAND_2$} and {NOR_2}.

To show that $NAND_2$ is functionally complete, all we have to do is show how to use $NAND_2$ to implement the functions in a set we have already seen is functionally complete. In particular, we will implement {AND_2, NOT} using only $NAND_2$.

NOT x can be implemented as $NAND_2$(x, x), which can also be written as x $NAND_2$ x. This can be verified using the truth table.

Once we have NOT implemented, then we can implement $AND_2(x, y)$ as NOT ($NAND_2(x, y)$). Negating $NAND_2$ gives us AND_2. We can show that NOR_2 is functionally complete by implementing OR_2 and NOT.

NAND and NOR

We only have to build the circuit using one kind of gate. Using a single gate is likely to make the circuit "larger", i.e., more gates than using many different kinds of gates, but usually it is better to use more of one kind of gate than fewer of many different gates.

1.4 Logic Gates and Truth Tables

The logic gates are the basic building blocks of any digital system. It is an electronic circuit having one or more than one input and only one output. Thus, the relationship between the input and the output is based on a certain logic. Based on this, the logic gates are named as AND gate or gate, NOT gate etc.

Three Representations of Logic Functions

Logical functions	AND	OR	NOT
Expression	X.Y	X+Y	X =X'

AND Gate

The circuit which performs an AND operation is shown in figure. This gate has n input ($n \geq 2$) and one output,

Y = A AND B AND C.... N

Y = A.B.C........ N

Y = ABC .... N

The AND gate is so named because, if 0 is called "false" then 1 is called "true," the gate acts in the same way as the logical "and" operator. The output is "true" when both inputs are "true." Otherwise, the output is "false."

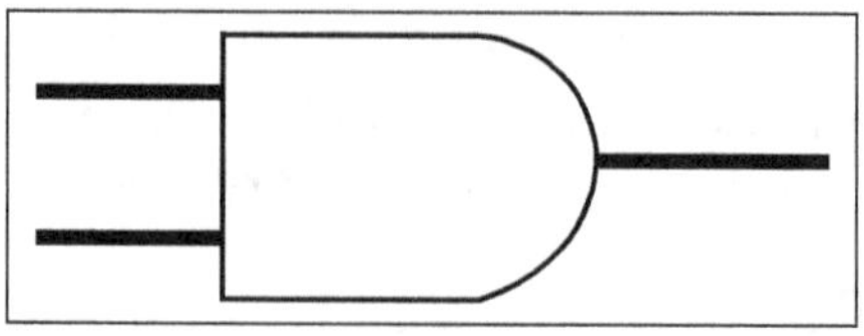

Symbol

Truth table:

Input 1	Input 2	Output
0	0	0
0	1	0
1	0	0
1	1	1

OR Gate

The circuit which performs an OR operation is shown in figure. This gate has n input (n ≥ 2) and one output,

Y = A OR B OR C .... N

$Y = A + B + C \ldots\ldots N$

The OR gate gets its name from the fact that it behaves like the logical inclusive "or." The output is "true" if either or both of the inputs are "true" and if both the inputs are "false," then the output is "false".

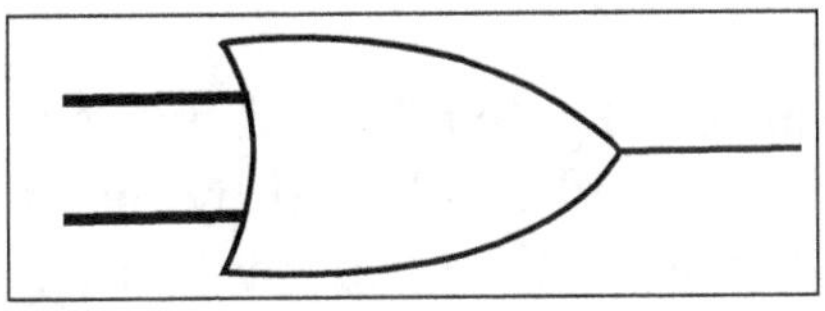

Symbol

Truth table:

Input 1	Input 2	Output
0	0	0
0	1	1
1	0	1
1	1	1

NOT Gate

The NOT gate is also known as Inverter. This gate has one input A and one output Y.

Y = NOT A

$Y = \overline{A}$

A logical inverter, sometimes called a NOT gate to differentiate it from other types of electronic inverter devices, has only one input. It reverses the logic state.

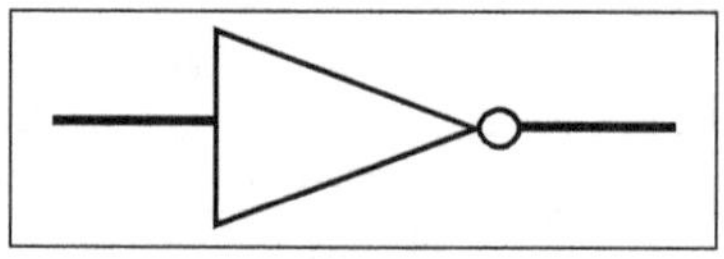

Symbol

Truth table:

Input	Output
1	0
0	1

1.4.1 Universal Logic Gates

NAND Gate

A NOT-AND operation is known as NAND operation. This gate has n input (n >= 2) and one output.

Y = A NOT AND B NOT AND C .... N

Y = A NAND B NAND C .... N

The NAND gate operates as an AND gate followed by a NOT gate. It acts in the similar manner of the logical operation "and", followed by negation. The output is "false" if both the inputs are "true". Otherwise, the output is "true".

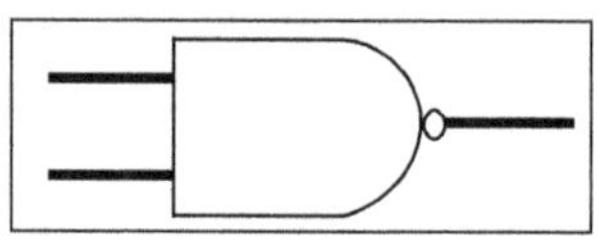

Symbol

Truth table:

Input 1	Input 2	Output
0	0	1
0	1	1
1	0	1
1	1	0

NOR Gate

The NOT-OR operation is known as NOR operation. This gate has n input (n >= 2) and one output.

Y = A NOT OR B NOT OR C .... N

Y = A NOR B NOR C .... N

The NOR gate is a combination of OR gate followed by an inverter. Its output is “true” if both the inputs are “false”. Otherwise, the output is “false”.

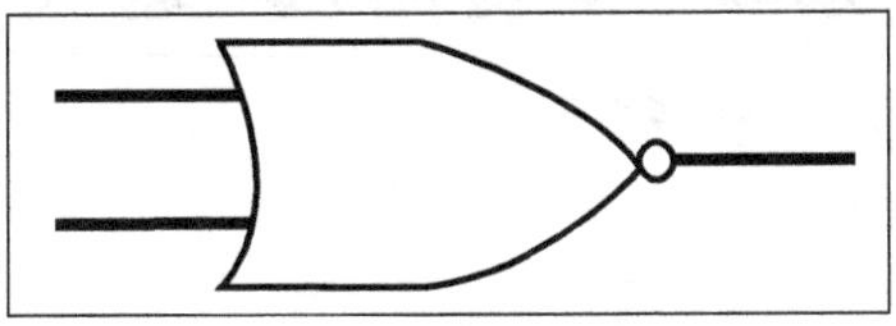

Symbol

Truth table:

Input 1	Input 2	Output
0	0	1
0	1	0
1	0	0
1	1	0

1.4.2 Algebraic Reduction and Realization Using Logic Gates

1. Let us reduce the expression $\overline{A+\overline{\overline{BC}}}\left(A\overline{B}+ABC\right)$.

Solution:

Given:

$$\overline{A+\overline{\overline{BC}}}\left(A\overline{B}+ABC\right)$$

$$\overline{A+\overline{\overline{BC}}}\left(A\overline{B}+ABC\right)=\overline{A}.BC\left(A\overline{B}+ABC\right)$$

$$=\overline{A}.BC.A\overline{B}+\overline{A}BC.ABC=0$$

2. Let us prove that (a) a + āb = a + b and (b) $\overline{xy}z+\overline{x}yz+x\overline{y}=\overline{x}z+x\overline{y}$

Solution:

$$\overline{xy}z+\overline{x}yz+x\overline{y}=\overline{x}z+x\overline{y}$$

(a) a + āb = a + b

$$a + \bar{a}b = a + ab + \bar{a}b = a + b(a + \bar{a})(\therefore a + \bar{a} = 1)$$

$$= a + b$$

L.H.S. = R.H.S.

(b) $\overline{xy}z+\overline{x}yz+x\overline{y}=\overline{x}z+x\overline{y}$

$$z\overline{x}\left(y+\overline{y}\right)+x\overline{y}=z\overline{x}\left(1\right)+x\overline{y}=z\overline{x}+x\overline{y}$$

L.H.S. = R.H.S.

3. Let us reduce the given expression using Boolean algebra.

$x'y'z' + x'y'z' + x'yz + xy'z + xyz$

Solution:

Given:

$x'y'z' + x'y'z' + x'yz + xy'z + xyz$

$= x'y'(z' + z) + yz(x' + x) + xy'z$

$= x'y' + yz + xy'z$

$= y'(x' + z) + yz$

$= y'x' + y'z + yz$

$= y'x' + z(y' + y)$

$= y'x' + z$

4. Let us reduce the given expression using Boolean algebra.

$abc' + ab'c + a'bc + abc$

Solution:

Given:

$abc' + ab'c + a'bc + abc$

$= ab(c' + c) + ab'c + a'bc$

$= ab + ab'c + a'bc = a(b + b'c) + a'bc$

$= ab + ac + a'bc = b(a + c) + ac = ab + ac + bc$

5. Let us reduce the given expression using Boolean algebra.

$p'q'r + p'qr' + p'qr + pqr' + pq'r'$

Solution:

$p'q'r + p'qr' + p'qr + pqr' + pq'r'$

$= p'r(q' + q) + qr'(p' + p) + pq'r'$

$= p'r + qr' + pq'r'$

$= r'(q + q'p) + p'r$

$= qr' + pr' + p'r$

$= qr' + p \oplus r$

1.5 Combinational Logic Design: Specifying the Problem

There are numerous practical applications of combinational logic circuits. It can be easily designed and built to control the outputs for various input combinations or to perform some mathematical or logical function.

Some circuits can be easily designed by directly writing a logical expression to solve the problem. For example, suppose that we desired to design a circuit to turn on a light using either of two switches. If either switch is up, the light will be lit. If both the switches are up or both switches are down, the light will not be lit. If we call the output OUT and inputs from the switches IN1 and IN2, we can fairly easily reason that $OUT = IN1 \cdot \overline{IN2} + \overline{IN1} \cdot IN2 = IN1 \oplus IN2$.

Combinational Logic Design Procedure:

- State the problem.
- Determine the required inputs and outputs.
- Assign variables to each input and output.
- Derive a truth table.
- Simplify output expressions.
- Implement each expression.

Example:

Let us design a circuit with 4 inputs that has outputs with a binary value equal to number of inputs that are HIGH.

Following the Combinational Logic Design Procedure above:

1. State the problem: Specified with the example in this case.

2. Determine the required number of inputs and outputs: 4 inputs are specified. If all 4 inputs are HIGH, the output code will be 100_2, so 3 output bits are required.

3. Assign variables to each input and output: Call the 4 inputs as A, B, C and D. Name the 3 outputs as E, F and G.

4. Truth table:

Inputs				outputs		
A	B	C	D	E	F	G
0	0	0	0	0	0	0
0	0	0	1	0	0	1
0	0	1	0	0	0	1
0	0	1	1	0	1	0
0	1	0	0	0	0	1
0	1	0	1	0	1	0
0	1	1	0	0	1	0
0	1	1	1	0	1	1
1	0	0	0	0	0	1
1	0	0	1	0	1	0
1	0	1	0	0	1	0
1	0	1	1	0	1	1
1	1	0	0	0	1	0
1	1	0	1	0	1	1
1	1	1	0	0	1	1
1	1	1	1	1	0	0

5. Simplify output expressions: K-maps for each output are shown below.

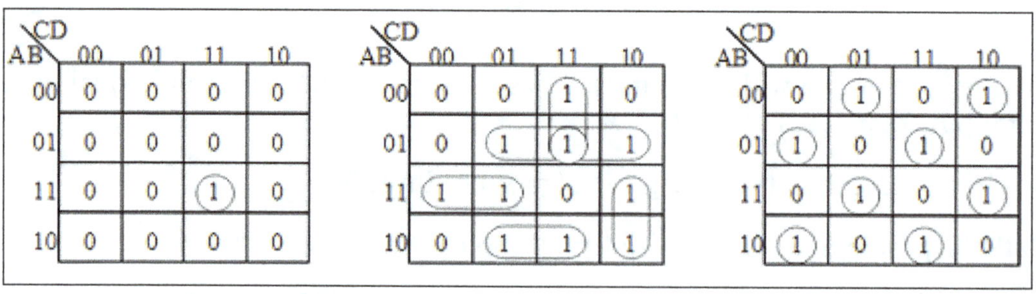

E = ABCD

F = A’CD + A’BD + A’BC + ABC’ + ACD’ + AB’D

$G = A \oplus B \oplus C \oplus D$

6. Implement each expression.

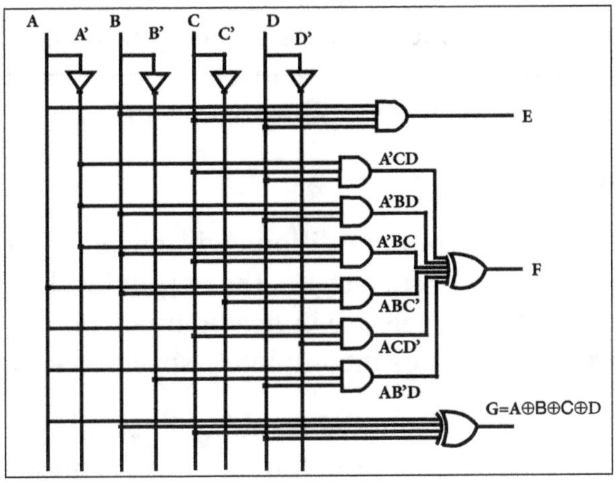

1.5.1 Canonical Logic Forms

The expanded form of Boolean expression, where each term contains all the Boolean variables in their true or complemented form, is known as the canonical form of the expression.

De Morgan has suggested two theorems that form important part of Boolean algebra. They are as follows:

- The complement of a product is equal to the sum of complements,

  (AB)' = A' + B'

- The complement of a sum term is equal to the product of the complements,

  (A + B)' = A'B'

Standard Form

Minterm

A product term containing all the variables of the function in either complemented or un complemented form is called as min term.

Maxterm

A sum term containing all the variables of the function in either complemented or un complemented form is called as max term.

Representing Standard SOP on K-Map

A Boolean expression in the sum of products form can be plotted on the Karnaugh map by placing 1 in each cell corresponding to a term in the sum of products expression. Remaining cells are filled with zeros.

Representing Standard POS on K-Map

A Boolean expression in the product of sums can be plotted on the Karnaugh map by placing 0 in each cell corresponding to a term in the expression. The remaining cells are filled with ones. This is described by the following examples.

Don't Care Condition

In some logic circuits, certain input conditions will never occur therefore the corresponding output never appears. In such cases, the output level is not defined, it may be either high or low. These output levels are indicated by 'X' or 'd' in the truth tables and are known as don't care conditions or incompletely specified functions.

For SOP, we enter 1's corresponding to the combinations of input variables which produce a high output and we enter 0's in the remaining cells of the K-map. In such situation, we have freedom to assume '0' or '1' as output for each of three combination in K-map. It is represented as X mark in the corresponding cell.

Sum of Products (SOP)

A Sum-Of-Products Boolean expression is literally a set of Boolean terms added together each term being a multiplicative combination of Boolean variables.

SOP-expression = term1 + term2 ... + term n

The product terms that include all of the input variables are called as minterms.

In a SOP-expression, we form a product of all the input variables (or their inverses) for each row of the truth table for which the result is logic 1. The output is the logical "sum" of these minterms.

The SOP-expressions are easy to generate by determining which rows of the table have an output of 1, writing the one product term for each row and then finally sum all the product terms. This creates a Boolean expression which is representing the truth table as a whole. The SOP-expressions lend themselves well to implementation as a set of AND gates (products) feeding into a single OR gate (sum).

Truth table:

A	B	C	W
0	0	0	0
0	0	1	1
0	1	0	1
0	1	1	1
1	0	0	0
1	0	1	1

1	1	0	1
1	1	1	0

The output can be expressed as,

$$W = \bar{A} \bullet \bar{B} \bullet C + \bar{A} \bullet B \bullet \bar{C} + \bar{A} \bullet B \bullet C + A \bullet \bar{B} \bullet C + A \bullet B \bullet \bar{C}$$

Product of Sums (POS)

An alternative to a Sum-Of-Products expression to account for all "high" (1) output conditions in the truth table is to generate the Product-Of-Sums or POS expression to account for all "low" (0) output conditions instead.

POS Boolean expressions may be generated from the truth tables quite easily, by determining the rows of the table which have an output of 0, writing the one sum term for each row and then finally multiply all the sum terms. This creates a Boolean expression which is representing the truth table as a whole.

These "sum" terms that include all of the input variables (or their inverses) are called maxterms. In the POS implementation, the output variable is the logical product of maxterms. Product-Of-Sums expressions lend themselves well for implementation as a set of OR gates (sums) feeding into a single AND gate (product).

Truth table:

A	B	C	Z
0	0	0	1
0	0	1	1
0	1	0	1
0	1	1	1
1	0	0	0
1	0	1	0
1	1	0	0
1	1	1	1

The output can be expressed as,

$$Z = (\bar{A} + B + C) \bullet (\bar{A} + B + \bar{C}) \bullet (\bar{A} + \bar{B} + C)$$

The Minterm and Maxterm Boolean expressions for the two-input OR gate are as follows,

$Y = A + B$ (maxterm or product-of-sums).

$Y = \bar{A}.B + A.\bar{B} + A.B$ (minterm of sum-of -products).

Truth Table:

A	B	Y
0	0	0
0	1	1
1	0	1
1	1	1

	$\overline{B}$	B
$\overline{A}$		1
A	1	1

Sum-of-Products K-Map.

	$\overline{B}$	B
$\overline{A}$		
A		1

Product-of-Sums K-Map.

Minterm Karnaugh map and Maxterm Karnaugh map of the 3 variable Boolean function,

$$Y = \overline{A}.\overline{B}.\overline{C} + \overline{A}.B.\overline{C} + A.\overline{B}.\overline{C} + A.B.\overline{C}$$

$$Y = \left(\overline{A} + \overline{B} + \overline{C}\right).\left(\overline{A} + B + \overline{C}\right).\left(A + \overline{B} + \overline{C}\right).\left(A + B + \overline{C}\right)$$

Truth Table:

A	B	C	Y
0	0	0	1
0	0	1	0
0	1	0	1
0	1	1	0
1	0	0	1
1	0	1	0
1	1	0	1
1	1	1	0

Sum-of-products K-map:

	$\overline{B}\,\overline{C}$	$\overline{B}\,C$	BC	$B\overline{C}$
$\overline{A}$	1			1
A	1			1

Products-of-sum K-map:

	$\overline{B}+\overline{C}$	$\overline{B}+C$	$B+C$	$B+\overline{C}$
$\overline{A}$	1			1
A	1			1

Minterm Karnaugh map and Maxterm Karnaugh map of the four variable Boolean function,

$$Y=\overline{A}.\overline{B}.\overline{C}.\overline{D}+\overline{A}.\overline{B}.\overline{C}.D+\overline{A}.B.\overline{C}.\overline{D}+\overline{A}.B.\overline{C}.D+A.\overline{B}.\overline{C}.\overline{D}+A.\overline{B}.\overline{C}.D+A.B.\overline{C}.\overline{D}+A.B.\overline{C}.D$$

$$Y=\left(A+B+\overline{C}+D\right).\left(A+B+\overline{C}+\overline{D}\right).\left(A+\overline{B}+\overline{C}+D\right).\left(A+\overline{B}+\overline{C}+D\right)$$

$$\left(\overline{A}+B+\overline{C}+D\right).\left(\overline{A}+B+\overline{C}+D\right).\left(\overline{A}+\overline{B}+\overline{C}+D\right).\left(\overline{A}+\overline{B}+\overline{C}+\overline{D}\right)$$

To illustrate the process of forming groups and then writing the corresponding minimized Boolean expression, the below figures respectively shows minterm and maxterm Karnaugh maps for the Boolean functions expressed by the below equations.

The minimized expressions as deduced from Karnaugh maps in the 2 cases are given by equation in the case of the minterm Karnaugh map and equation in the case of the maxterm Karnaugh map,

$$Y=\overline{A}.\overline{B}.\overline{C}.\overline{D}+\overline{A}.\overline{B}.C.\overline{D}+\overline{A}.B.\overline{C}.D+\overline{A}.B.C.D+A.\overline{B}.\overline{C}.\overline{D}+A.\overline{B}.C.\overline{D}+A.B.\overline{C}.D+A.B.C.D$$

$$Y=\left(A+B+C+\overline{D}\right).\left(A+B+\overline{C}+\overline{D}\right).\left(A+\overline{B}+C+D\right).\left(A+\overline{B}+C+\overline{D}\right).\left(A+\overline{B}+\overline{C}+\overline{D}\right)$$

$$\left(A+\overline{B}+\overline{C}+D\right).\left(\overline{A}+\overline{B}+C+\overline{D}\right).\left(\overline{A}+\overline{B}+\overline{C}+\overline{D}\right).\left(\overline{A}+B+C+\overline{D}\right).\left(\overline{A}+B+\overline{C}+\overline{D}\right)$$

$$Y=\overline{B}.\overline{D}+B.D$$

$$Y=\overline{D}.\left(A+\overline{B}\right)$$

Truth Table:

A	B	C	D	Y
0	0	0	0	1
0	0	0	1	1
0	0	1	0	0
0	0	1	1	0
0	1	0	0	1
0	1	0	1	1

0	1	1	0	0
0	1	1	1	0
1	0	0	0	1
1	0	0	1	1
1	0	1	0	0
1	0	1	1	0
1	1	0	0	1
1	1	0	1	1
1	1	1	0	0
1	1	1	1	0

Sum of products K-map:

	$\overline{C}\,\overline{D}$	$\overline{C}D$	CD	$C\overline{D}$
$\overline{A}\,\overline{B}$	1	1		
$\overline{A}B$	1	1		
AB	1	1		
$A\overline{B}$	1	1		

Product of sums K-map:

	$\overline{C}+\overline{D}$	$\overline{C}+D$	C+D	$C+\overline{D}$
$\overline{A}+\overline{B}$	1	1		
$\overline{A}+B$	1	1		
A + B	1	1		
$A+\overline{B}$	1	1		

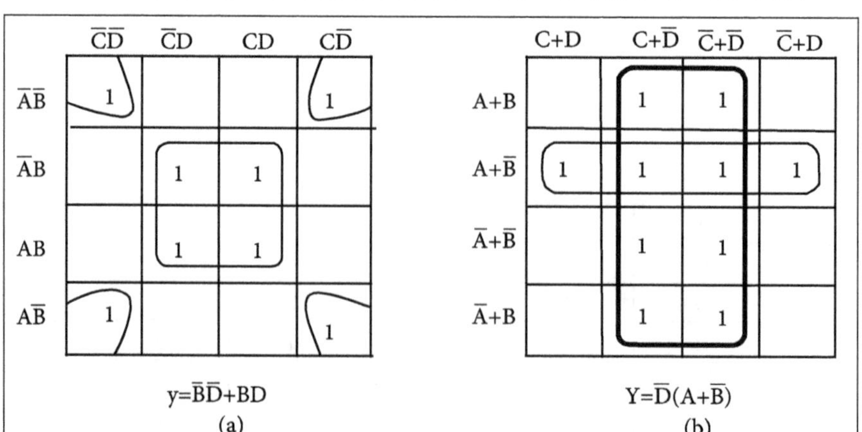

Group formation in minterm and maxterm Karnaugh maps.

Problems

1. Let us prove that a function $F = A + \bar{B}C$ expressed as a sum of its minterms is equivalent to a function expressed as a product of its maximum terms.

Solution:

Given:

$$F = A + \bar{B}C$$

SOP,

$$\begin{aligned} F &= A + \bar{B}C \\ &= A(B+\bar{B})(C+\bar{C}) + \bar{B}C(A+\bar{A}) \\ &= (AB + A\bar{B})(C+\bar{C}) + A\bar{B}C + \bar{A}\bar{B}C \end{aligned}$$

$$\begin{aligned} F &= ABC + A\bar{B}C + ABC\bar{} + A\bar{B}\bar{C} + A\bar{B}C + \bar{A}\bar{B}C \\ &= ABC + AB\bar{C} + A\bar{B}C + A\bar{B}\bar{C} + \bar{A}\bar{B}C \\ &= M_7 + M_6 + M_5 + M_4 + M_1 \\ F &= \sum(1,4,5,6,7) \end{aligned}$$

2. Let us show that a function $F = A + \bar{B}C$ expressed as a sum of its minterms is equivalent to a function expressed as a product of its maximum terms.

Solution:

$$F = A + \bar{B}C$$

Given:

POS,

$$\begin{aligned} F &= A + \bar{B}C \\ &= (A+\bar{B})(A+C) = (A+\bar{B}+C\bar{C})(A+C+B\bar{B}) \\ &= (A+\bar{B}+C)(A+\bar{B}+\bar{C})(A+B+C) \\ &= M_{2,} M_{3,} M_0 \\ F &= \pi(0,2,3) \end{aligned}$$

3. Let us express the function $F = A + \bar{B}C$ in Canonical SOP form.

Solution:

Given:

$$F = A + \bar{B}C$$

Sum of Product (SOP),

$$Y = A + \bar{B}.C = A(B + \bar{B}) + \bar{B}C(A + \bar{A}) = AB + A\bar{B} + A\bar{B}C + \bar{A}\bar{B}C$$
$$= AB(C + \bar{C}) + A\bar{B}(C + \bar{C}) + A\bar{B}C + \bar{A}\bar{B}C$$
$$= ABC + AB\bar{C} + A\bar{B}C + A\bar{B}\bar{C} + A\bar{B}C + \bar{A}\bar{B}C$$
$$Y = m_7 + m_6 + m_5 + m_3 + m_5 + m_1$$
$$Y = \sum(1, 4, 5, 6, 7)$$

4. Here we see an another example to express the function in Canonical POS form.

Solution:

Given:

$$F = A + \bar{B}C$$

Product of Sum (POS),

$$Y = A + \bar{B}C = (A + \bar{B})(A + C)[\therefore A + BC = (A + B)(A + C)]$$
$$= (A + \bar{B} + C)(A + \bar{B} + \bar{C})(A + B + C)(A + B + C)$$
$$= (A + \bar{B} + C).(A + \bar{B} + \bar{C}).(A + B + C)$$
$$Y = M_2 M_3\ M_0 = M_0\ M_2\ M_3$$
$$Y = \pi(0, 2, 3)$$

5. Here let us minimize and implement the following multiple output functions in SOP form.

$$f_1 = \sum m\ (0,2,6,10,11,12,13) + d\ (3,4,5,14,15)$$
$$f_2 = \sum m\ (1,2,6,7,8,13,14,15) + d\ (3,5,12)$$

Solution:

Given:

$$f_1 = \sum m\ (0,2,6,10,11,12,13) + d\ (3,4,5,14,15)$$
$$f_2 = \sum m\ (1,2,6,7,8,13,14,15) + d\ (3,5,12)$$

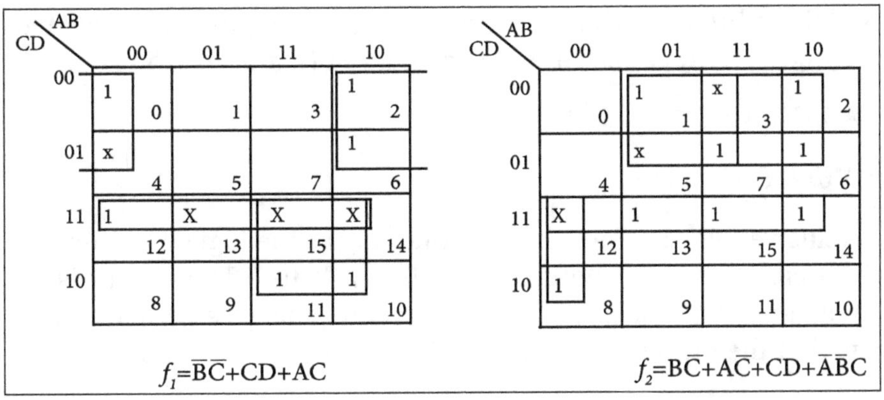

The K maps are filled with ones and don't cares using the expression. After reduction, we find that CD occurs both in f_1 and f_2. So it can be shared.

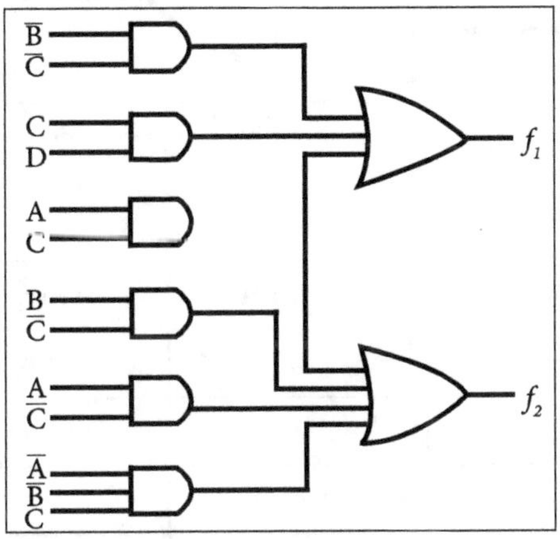

Logic Circuit Implementation.

6. Here let us convert the given expression in canonical SOP form Y = AC + AB + BC.

Solution:

Given:

$$Y = AC + AB + BC$$

$$= AC(B + B') + AB(C + C') + (A + A')BC$$

$$= ABC + ABC' + AB'C + AB'C' + ABC + ABC' + ABC$$

$$= ABC + ABC' + AB'C + AB'C' \quad [A + A = 1]$$

1.5.2 EX-OR Equivalence Operations

XOR Gate

The XOR or Ex-OR gate is a special type of gate. This gate can be used in the half adder, full adder and the subtractor. Here, the exclusive-OR gate can be abbreviated as EX-OR gate or sometime as XOR gate. This gate has n input (n ≥ 2) and one output,

Y = A XOR B XOR C .... N

$Y = A \oplus B \oplus C \ldots N$

The XOR (exclusive-OR) gate acts in the same way as the logical "either/or." The output is "true" if either, but not both the inputs are "true." The output is "false" when both the inputs are "true" or if both the inputs are "false". Another way of analyzing this circuit is that the output is 1 when the inputs are different but 0 if the inputs are same.

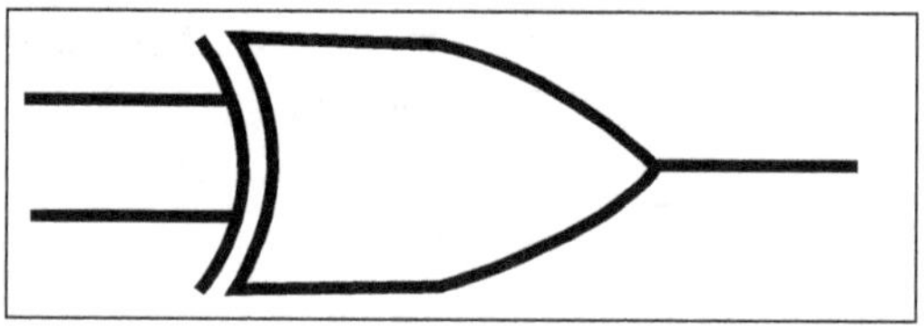

Symbol

Truth table:

Input 1	Input 2	Output
0	0	0
0	1	1
1	0	1
1	1	0

XNOR Gate

The XNOR gate is a special type of gate. It can be used in the half adder, full adder and subtractor. The exclusive-NOR gate is shortly known as EX-NOR gate or sometime as XNOR gate. It has n input (n ≥ 2) and one output.

Y = A XNOR B XNOR C .... N

The XNOR (exclusive-NOR) gate is a combination XOR gate followed by an inverter. Its output is "true" if the inputs are the same and "false" if the inputs are different.

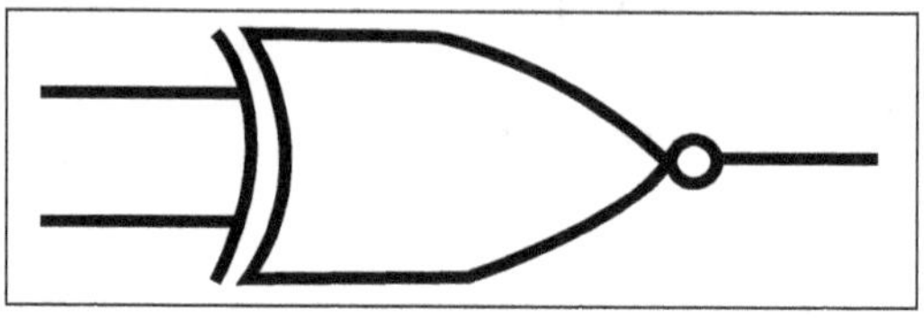

Symbol

Truth table:

Input 1	Input 2	Output
0	0	1
0	1	0
1	0	0
1	1	1

1.5.3 Logic Array

The logic gates used can be two level AND-OR, NAND-NAND or NOR-NOR configuration. The PLA consist of two level AND-OR circuit on the same chip. The AND matrix can be used to implement the product terms in the SOP form and the OR matrix is used for implementing the sum of product terms.

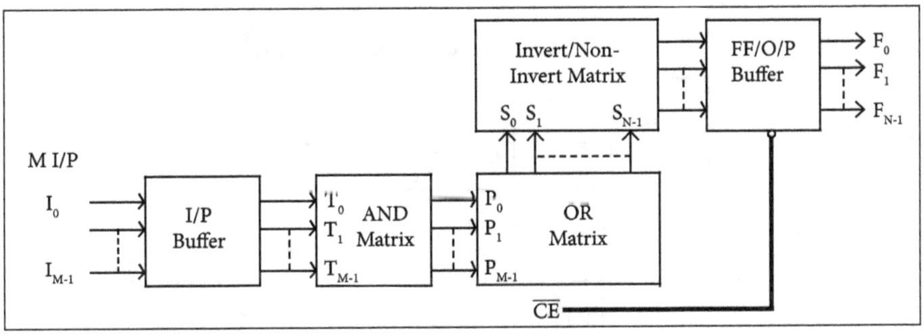

PLA

1.6 K-Maps: Two, Three and Four Variable K-Maps

A Karnaugh map (or) K-map is a pictorial form of truth table in which the map diagram is made up of cells with each cell representing one minterm or maxterm of the function. It provides a simple straight forward procedure for minimizing Boolean function.

Construction of a Karnaugh Map

An n-variable Karnaugh map has 2n squares and each possible input is allocated a square. In the case of a minterm Karnaugh map, '1' is placed in all those squares for which the output is '1' and '0' is placed in all those squares for which the output is '0'. 0s are omitted for simplicity. An 'X' is placed in squares corresponding to don't care' conditions.

In the case of a maxterm Karnaugh map, a '1' is placed in all those squares for which the result is '0' and a '0' is placed for input entries corresponding to a '1' result. Then, 0s are omitted for simplicity and an 'X' is placed in squares corresponding to don't care' conditions. The choice of terms identifying different rows and columns of a Karnaugh map is not unique for a given number of variables.

The only condition to be satisfied is that the designation of adjacent rows and adjacent columns should be the same except for one of the literals being complemented. Also, the extreme rows and extreme columns are considered adjacent. Some of the possible designation styles for 2, 3 and 4 variable minterm, Karnaugh maps are shown in the figure below.

Two variable K-Map:

	$\overline{B}$	B
$\overline{A}$		
A		

Three variable K-Map:

	$\overline{B}\,\overline{C}$	$\overline{B}\,C$	BC	$B\overline{C}$
$\overline{A}$				
A				

Four variable K-Map:

	$\overline{C}\,\overline{D}$	$\overline{C}\,D$	CD	$C\overline{D}$
$\overline{A}\,\overline{B}$				
$\overline{A}\,B$				
AB				
$A\overline{B}$				

A = 0

BC \ DE	$\overline{D}\,\overline{E}$ 00	$\overline{D}E$ 01	DE 11	$D\overline{E}$ 10
$\overline{B}\,\overline{C}$ 00	0	1	3	2
$\overline{B}C$ 01	4	5	7	6
BC 11	12	13	15	14
$B\overline{C}$ 10	8	9	11	10

A = 1

BC \ DE	$\overline{D}\,\overline{E}$ 00	$\overline{D}E$ 01	DE 11	$D\overline{E}$ 10
$\overline{B}\,\overline{C}$ 00	16	17	19	18
$\overline{B}C$ 01	20	21	23	22
BC 11	28	29	31	30
$B\overline{C}$ 10	24	25	27	26

Five variable K-Map.

$\bar{A}\bar{B}$

CD \ EF	$\bar{E}\bar{F}$ 00	$\bar{E}F$ 01	EF 11	$E\bar{F}$ 10
$\bar{C}\bar{D}$ 00	0	1	3	2
$\bar{C}D$ 01	4	5	7	6
CD 11	12	13	15	14
$C\bar{D}$ 10	8	9	11	10

$\bar{A}B$

CD \ EF	$\bar{E}\bar{F}$ 00	$\bar{E}F$ 01	EF 11	$E\bar{F}$ 10
$\bar{C}\bar{D}$ 00	16	17	19	18
$\bar{C}D$ 01	20	21	23	22
CD 11	28	29	31	30
$C\bar{D}$ 10	24	25	27	26

$A\bar{B}$

CD \ EF	$\bar{E}\bar{F}$ 00	$\bar{E}F$ 01	EF 11	$E\bar{F}$ 10
$\bar{C}\bar{D}$ 00	32	33	35	34
$\bar{C}D$ 01	36	37	39	38
CD 11	44	45	47	46
$C\bar{D}$ 10	40	41	43	42

AB

CD \ EF	$\bar{E}\bar{F}$ 00	$\bar{E}F$ 01	EF 11	$E\bar{F}$ 10
$\bar{C}\bar{D}$ 00	48	49	51	50
$\bar{C}D$ 01	52	53	55	54
CD 11	60	61	63	62
$C\bar{D}$ 10	58	59	61	60

Six variable K-Map.

The style of row identification need not be the same as that of column identification as long as it meets the basic requirement with respect to adjacent terms. It is, however, accepted practice to adopt a uniform style of row and column identification. Also, the style shown in the figure below is more commonly used. Having drawn the Karnaugh map, the next step is to form groups of 1s as per the subsequent guidelines:

- Each square containing a '1' must be considered at least once, although it can be considered as often as desired.
- The goal should be to account for all the marked squares in the minimum number of groups.
- The number of squares in a group should always be a power of 2, i.e. groups can have 1, 2, 4, 8, 16, squares.

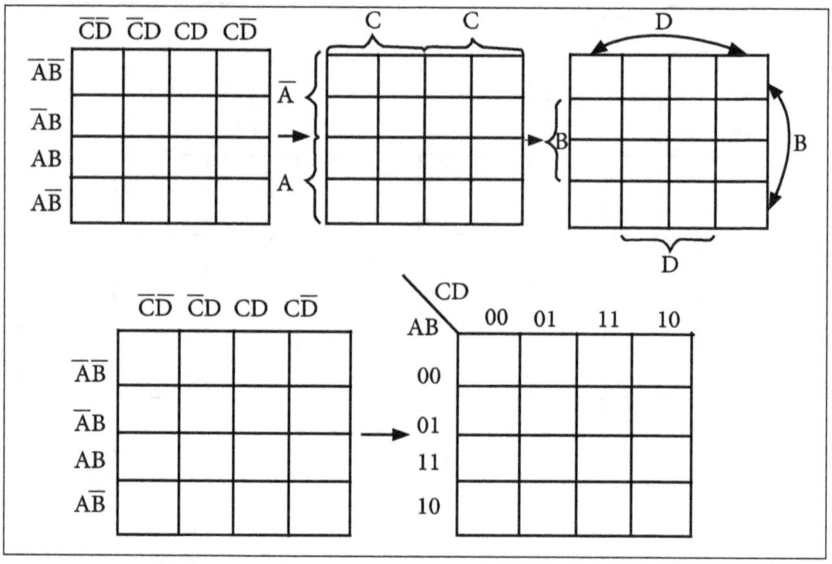

Different Styles of row and column identification.

- Each group should be as large as possible, which means that a square must not be accounted for by itself if it can be accounted for by a group of 2 squares. A group of two squares should not be made if the involved squares can be included in a group of four squares and so on.
- Don‘t care entries can be used in accounting for all of 1-squares to make optimum groups. They are marked 'X' in the corresponding squares. It is, however, not necessary to account for all don‘t care‘ entries.

Having accounted for groups with all 1s, the minimum sum-of-products or product-of-sums expressions can be written directly from the Karnaugh map. Minterm Karnaugh map and Maxterm Karnaugh map of the Boolean function of a 2-input OR gate.

Limitations of Karnaugh Map

- The map method is restricted in its capability because they are useful for simplifying only the Boolean expression represented in standard form.
- Generally it is limited to six variable map (i.e.) more than six variables involving expressions are not reduced.

Problems

1. Let us solve $g(W, X, Y, Z) = \sum m\ (1, 3, 4, 6, 11) + \sum d\ (0, 8, 10, 12, 13)$.

Solution:

Given:

$$g(W, X, Y, Z) = \sum m\ (1, 3, 4, 6, 11) + \sum d\ (0, 8, 10, 12, 13)$$

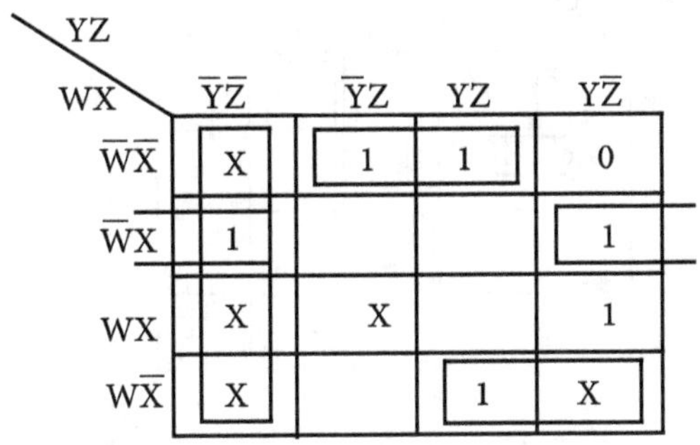

$Y=\bar{Y}\bar{Z}+\bar{W}\bar{Y}Z+\bar{W}X\bar{Z}+X\bar{W}Y$

2. Let us simplify the following function using K - map and also implement the function using logic gates f(A, B, C) = π(0, 4, 6).

Solution:

Given:

$$f(A, B, C) = \pi(0, 4, 6)$$

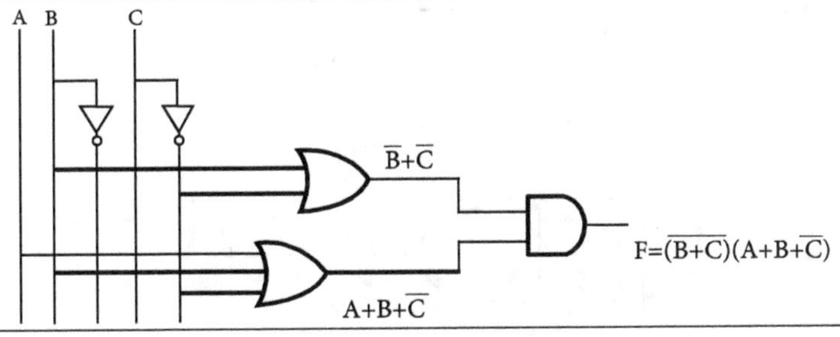

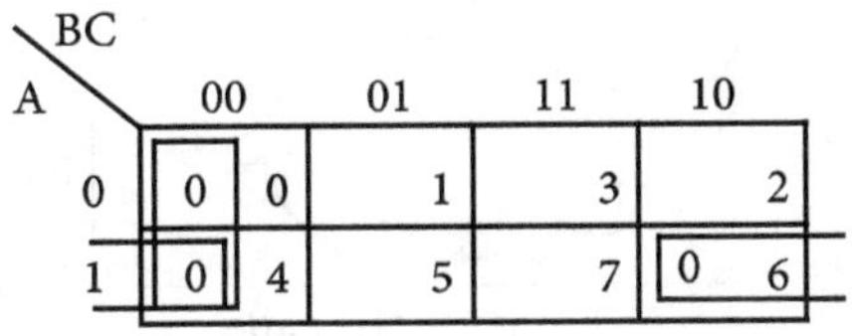

$F=(\overline{B+C})(A+B+\bar{C})$

3. Let us reduce the following expression using K-map: $F = m_2 + m_3 + m_4 + m_6 + m_7 + m_9 + m_{11} + m_{12}$.

Solution:

Given:

$$F = m_2 + m_3 + m_4 + m_6 + m_7 + m_9 + m_{11} + m_{12}$$

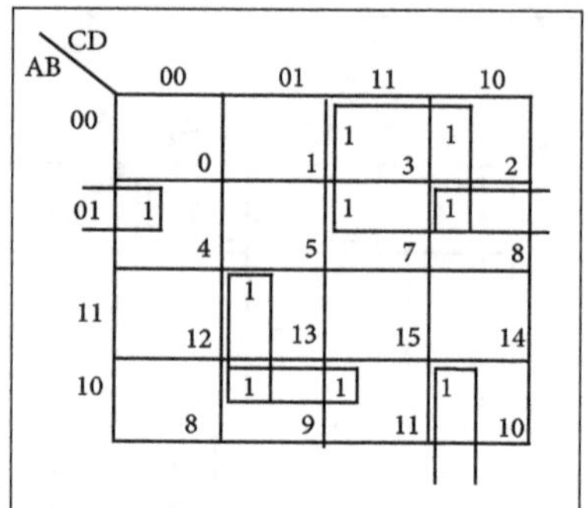

$F=\bar{A}B\bar{D}+A\bar{C}D+A\bar{B}D+\bar{A}C$

1.6.1 NAND and NOR Logic Implementations

The NAND and the NOR gates are generally faster and use fewer components than AND or OR gates. It can be implemented by using only NAND gates or only NOR gates with any logic function. NAND gate or NOR gate forms a functionally complete set where any switching function can be expressed in terms of NAND gate or NOR gate. Similarly, the set AND, OR and NOT shall form a functionally complete set since any Boolean function can be expressed in terms of SOP and POS only by using the AND, OR and NOT operations.

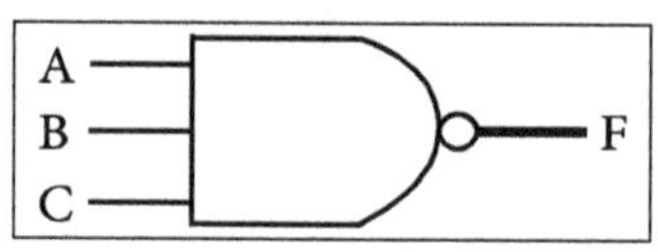

3 - Input NAND gate.

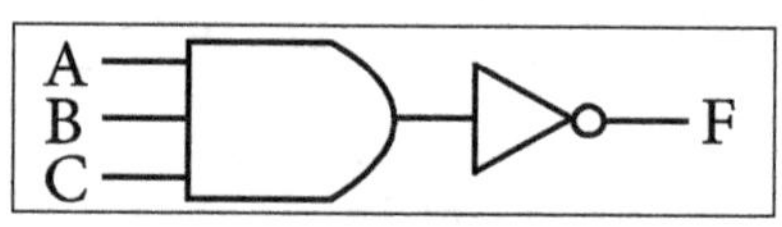

NAND gate equivalent.

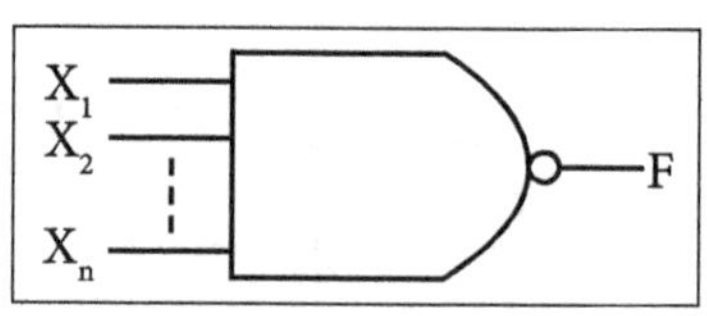

n-input NAND gate.

N-Input Nand Gate

$$F=(X_1 X_2 .....X_n)' = X_1' + X_2' + ... + X_n'$$

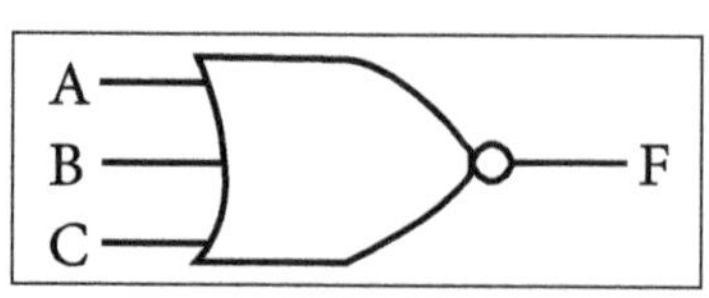

3 - Input NOR gate.

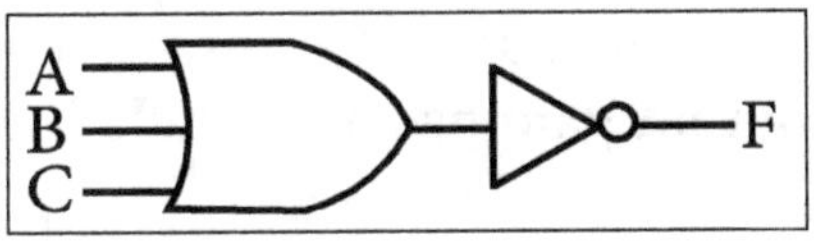

NOR gate equivalent.

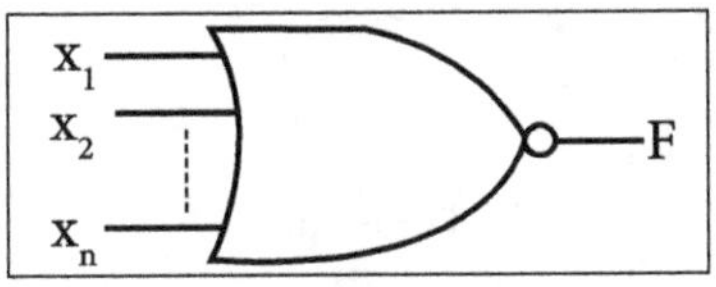

N-input NOR gate.

n-input NOR gate,

$$F = (X_1 + X_2 + ... + X_n)' = X_1' X_2' .... X_n'$$

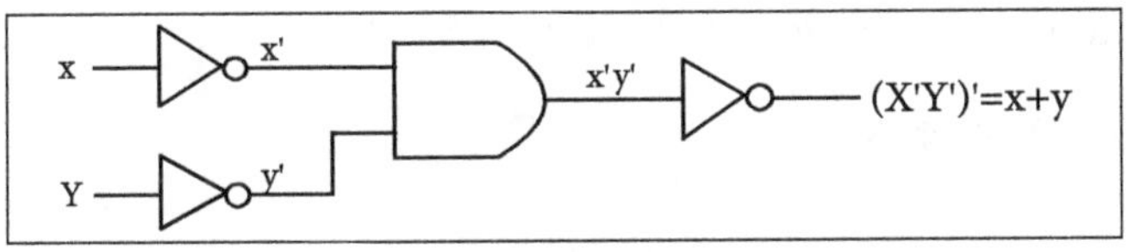

NAND Gate Realization of NOT and OR.

Design of 2-Level Circuits using NAND and NOR Gates

- Find a minimum SOP expression for F.
- Then, draw the corresponding 2-level AND-OR circuit.
- Replace all gates with NAND gates leaving the gate interconnections unchanged. When the output gate has any single literals as inputs, complement these literals.

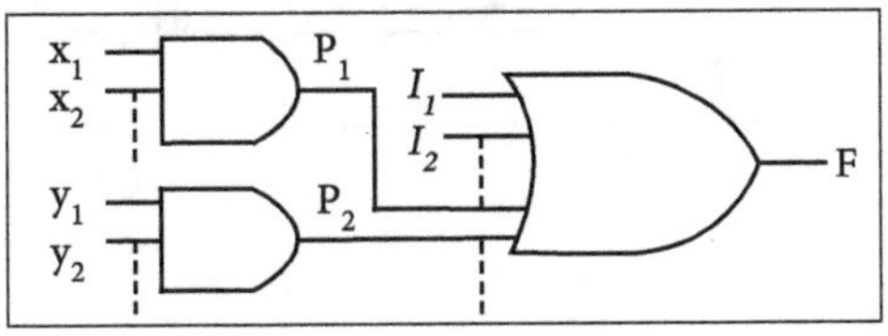

Before transmission.

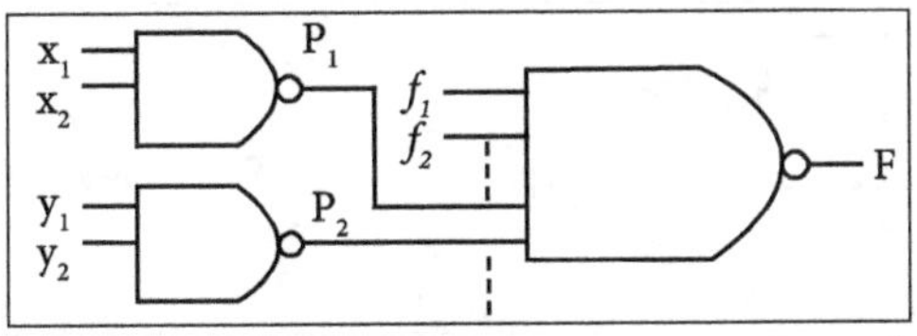

After transmission. AND-OR to NAND-NAND Transformation.

Steps for Designing a Minimum 2-Level Nor-Nor Circuit

- Find a minimum POS expression for F.
- Draw the corresponding 2-level OR-AND circuit.

- Then, replace all gates with NOR gates leaving the gate interconnections unchanged. When the output gate has any single literals as inputs and complement these literals.

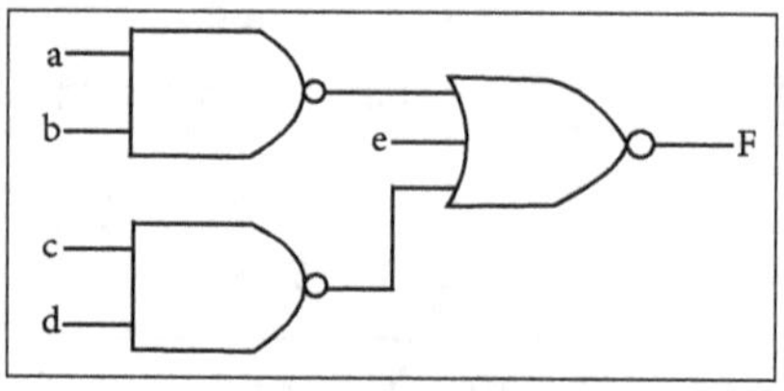

NOR-NOR Transformation.

F = [(ab)' + (cd)' + e]' = abcde'

Problems

1. Let us see the NAND gate implementation.

F(x, y, z) = (1, 2, 3, 4, 5, 7)

Solution:

Given:

F(x, y, z) = (1, 2, 3, 4, 5, 7)

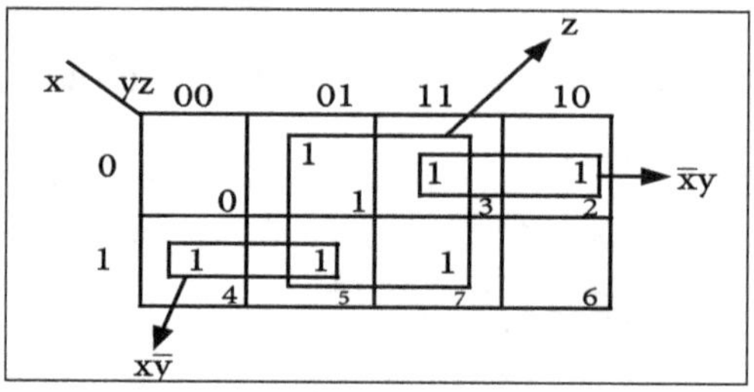

$F = x\bar{y} + \bar{x}y + z$

NAND Implementation

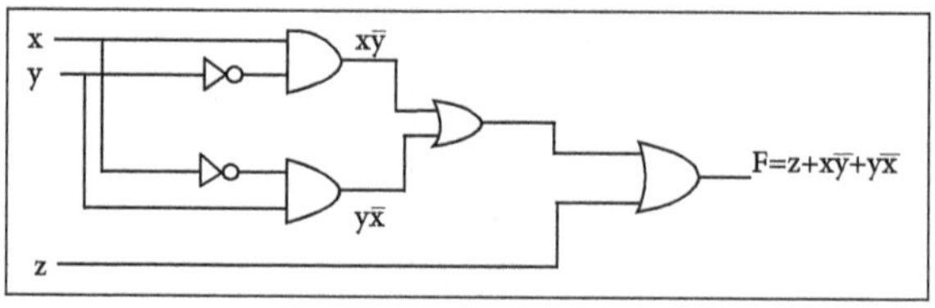

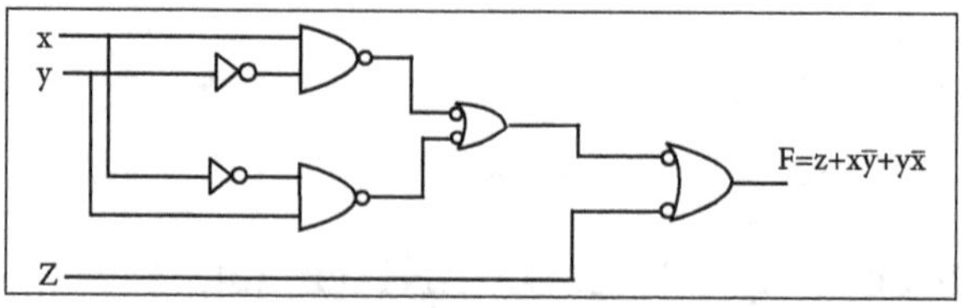

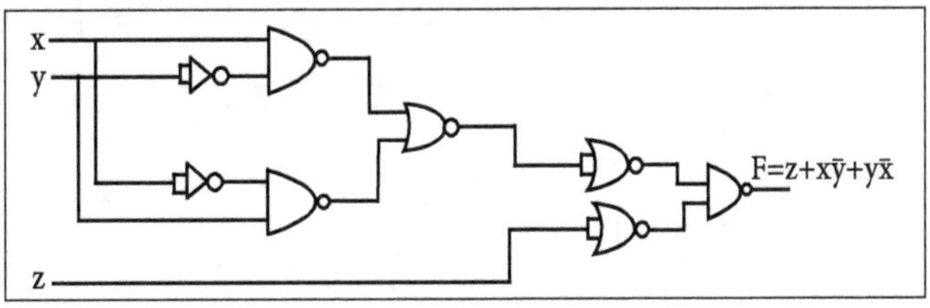

2. Let us draw the NAND logic diagram that implement the complement of the functions $F(A, B, C, D) = \sum(0, 1, 2, 3, 4, 8, 9, 12)$.

Solution:

Given:

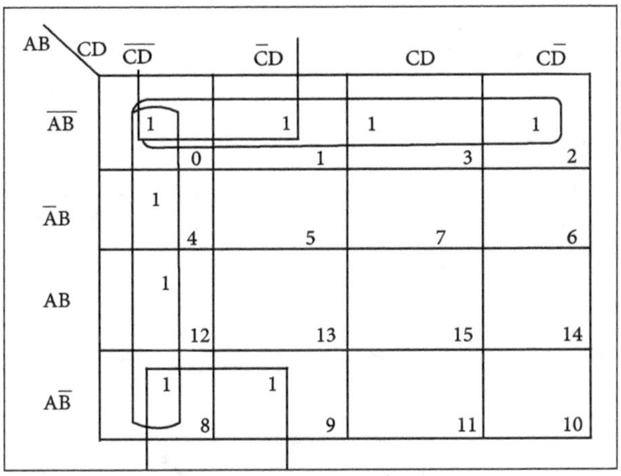

$$F(A, B, C, D) = \sum(0, 1, 2, 3, 4, 8,9, 12)$$

$$F = \overline{B}\overline{C}\overline{D} + \overline{B}\overline{C} + \overline{A}\overline{B}$$

$$F = \overline{\overline{C}\overline{D}} + \overline{\overline{B}\overline{C}} + \overline{\overline{A}\overline{B}}$$

$$= \left(\overline{\overline{C}\overline{D}}\right)\left(\overline{\overline{B}\overline{C}}\right)\left(\overline{\overline{A}\overline{B}}\right)$$

$$= (C+D)(B+C)(A+B)$$

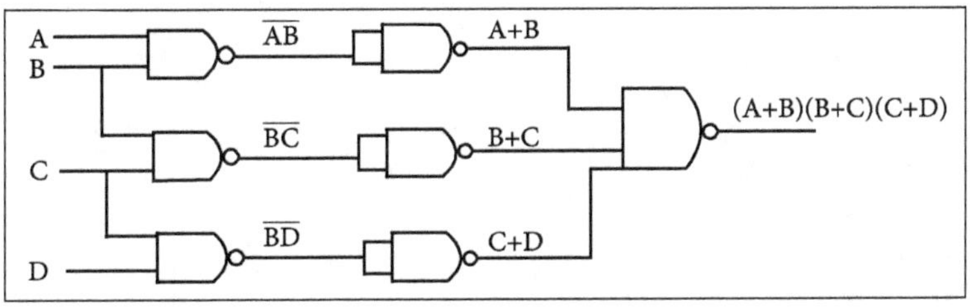

3. Let us see the implementation of EXOR using NAND and NOR gate.

Solution:

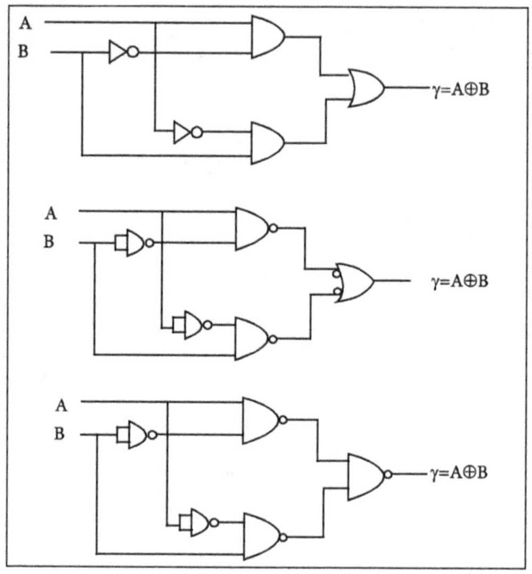

EXOR using NAND

EXOR using NOR

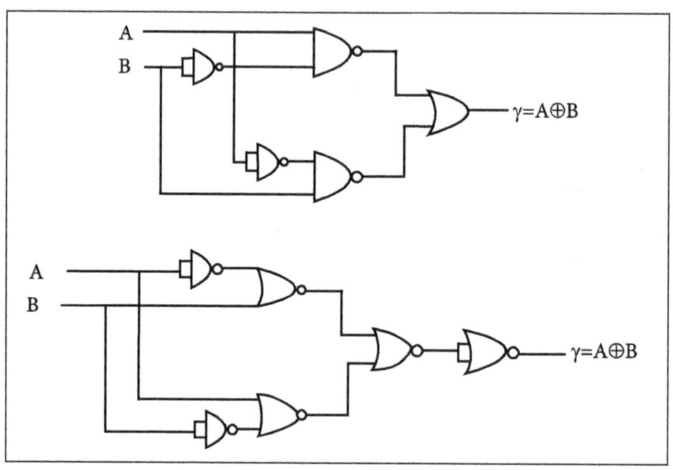

2

Logic Components, Synchronous Sequential Logic Design and Binary Counters

2.1 Logic Components: Concept of Digital Components

A simple processor illustrates many of the basic components used in any digital system.

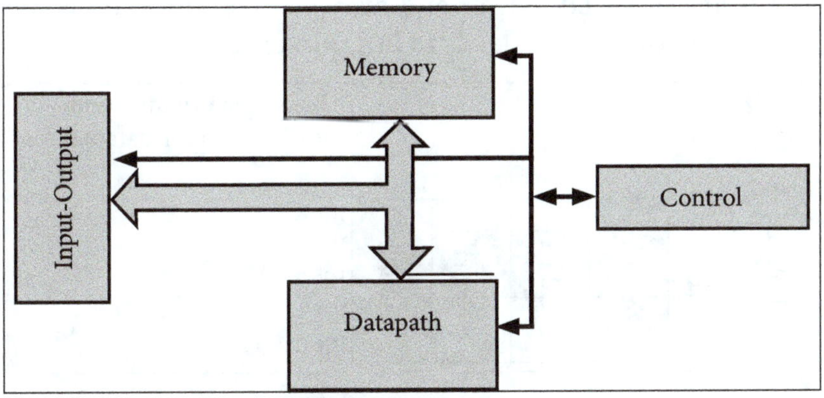

Basic components.

Data-Path

The core and all other components are support units that store either the results of the data-path or determine what happens in the next cycle.

Memory

A broad range of classes exist determined by the way data is accessed:

- Read-Only vs Read-Write.
- Sequential vs Random access.
- Single-ported vs Multi-ported access.
- Or by their data retention characteristics.
- Dynamic vs Static.

Control

A FSM (sequential circuit) implemented using random logic, PLA's or memories.

Interconnect and Input-Output

- Parasitic resistance, capacitance and inductance affect performance of wires both on and off the chip.
- Growing die size increases the length of the on-chip interconnect, increasing the value of the parasitics.

Data-path elements include adders, multipliers, shifters, BFU's, etc.

The speed of these elements often dominates the overall system performance, so optimization techniques are important. However, the task is non-trivial since there are multiple equivalent logic and circuit topologies to choose from, each with advantage/ disadvantage in terms of speed, power and area. Also, optimizations focused at one design level, e.g., sizing transistors, leads to inferior designs.

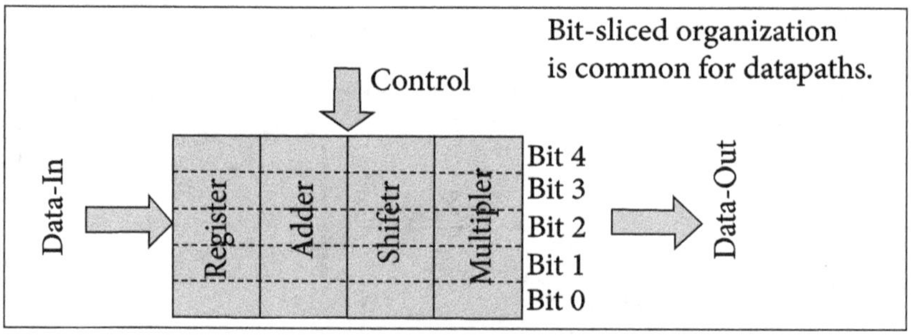

2.1.1 Binary Adders, Subtraction and Multiplication

Half Adder

A logic circuit which performs the addition of two bits is a half adder. A half-adder is an arithmetic circuit block which can be used to add two bits. Such a circuit thus has two inputs that represent the two bits to be added and two outputs with one which produces the SUM output and the other producing the CARRY. The figure shows the truth table of a half adder that shows all possible input combinations and the corresponding outputs.

The Boolean expressions for the SUM and CARRY outputs are shown in the equations given below,

$$\text{SUM } S = A.\overline{B} + \overline{A}.B$$

$$\text{CARRY } C = A.B$$

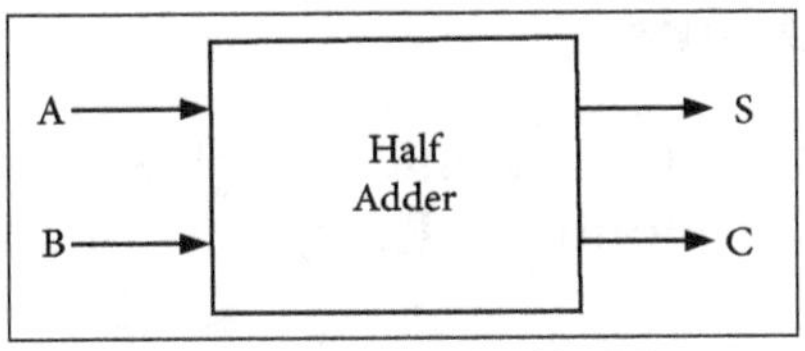

Block diagram of half adder.

Truth table:

A	B	S	C
0	0	0	0
0	1	1	0
1	0	1	0
1	1	0	1

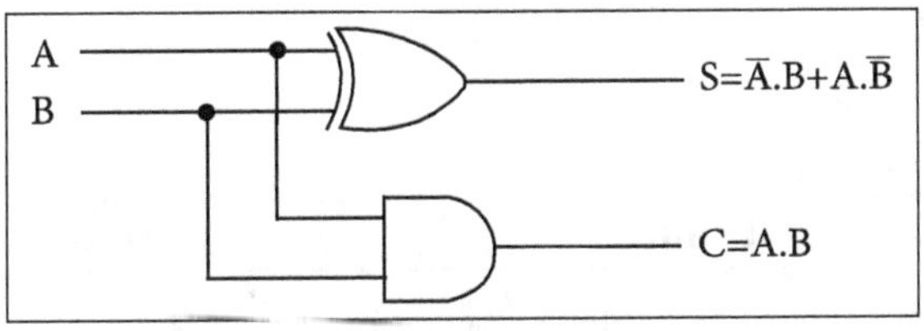

Half adder.

Full Adder

The circuit that performs the addition of three bits is a full adder.

Implementation of a Full Adder using Half Adders

Full adder accepts three inputs and it generates two outputs.

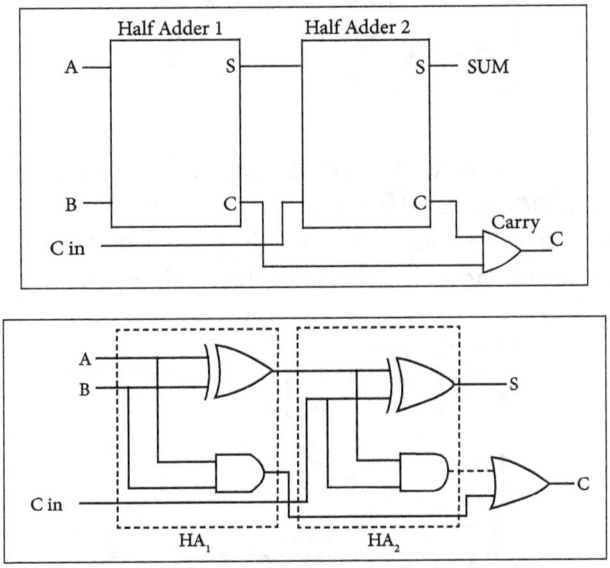

Implementation of Full Adder using half adders.

Applications of Full Adders

- Full adder reduces the circuit complexity. This is used to construct a ripple carry counter to add n-bit numbers. Thus, it is used in the Arithmetic and Logic Unit also.
- This can be used in Processor chip such as Snapdragon, Intel Pentium or Exynous for CPU part which consists of ALU. Therefore, this block is used to make operations such as Add, subtract, Multiply etc.
- A full adder adds the binary numbers and accounts for values carried in as well as out.
- A one-bit full adder adds three one-bit numbers, often written as A, B and C where A and B are the operands and Cin is a bit carried in from the previous less significant stage.

Truth Table:

Input			**Output**	
A	B	C_{in}	S	C
0	0	0	0	0
0	0	1	1	0
0	1	0	1	0
0	1	1	0	1
1	0	0	1	0
1	0	1	0	1
1	1	0	0	1
1	1	1	1	1

Half Subtractor

A half-subtractor is a combinational circuit that can be used to subtract one binary digit from another to produce a DIFFERENCE output and a BORROW output. The BORROW output here specifies whether a = 1 has been borrowed to perform the subtraction. The Boolean expressions for the two outputs are given by the equations.

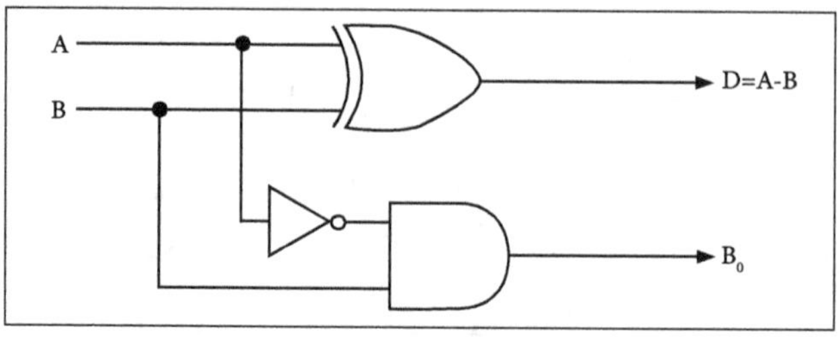

Half subtractor.

Full Subtractor

A full subtractor performs subtraction operation on two bits such as a minuend and a subtrahend and also takes into consideration whether a=1 has already been borrowed by the previous adjacent lower minuend bit or not. Finally, as a result there are three bits to be handled at the input of a full subtractor namely the two bits to be subtracted and a borrow bit that is designated as Bin. There are two outputs namely the DIFFERENCE output D and a BORROW output B_o.

The BORROW output bit tells whether the minuend bit needs to borrow a = 1' from the next possible higher minuend bit. The figure below shows the truth table of a full subtractor. The Boolean expressions are given by the equations for two output variables,

$$D = \overline{A}.\overline{B}.B_{in} + \overline{A}.B.\overline{B}_{in} + A.\overline{B}.\overline{B}_{in} + A.B.B_{in}$$

$$B_o = \overline{A}.\overline{B}.B_{in} + \overline{A}.B.\overline{B}_{in} + \overline{A}.B.B_{in} + A.B.B_{in}$$

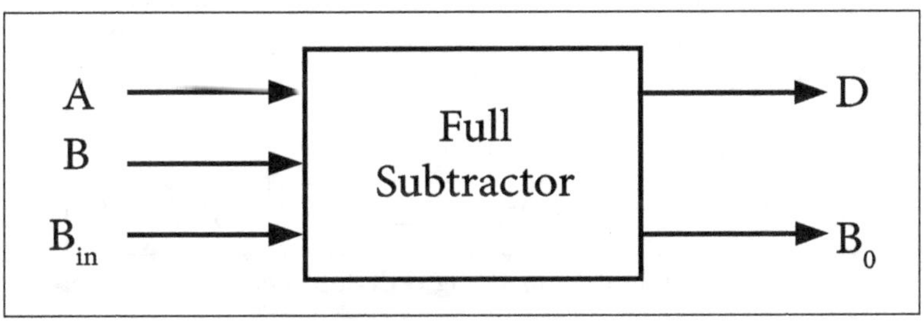

Block diagram of full subtractor.

Truth Table:

Minuend (A)	Subtrahend (B)	Borrow In (B_{in})	Difference (D)	Borrow out (B_o)
0	0	0	0	0
0	0	1	1	1
0	1	0	1	1
0	1	1	0	1
1	0	0	1	0
1	0	1	0	0
1	1	0	0	0
1	1	1	1	1

K Maps for Difference and Borrow Outputs

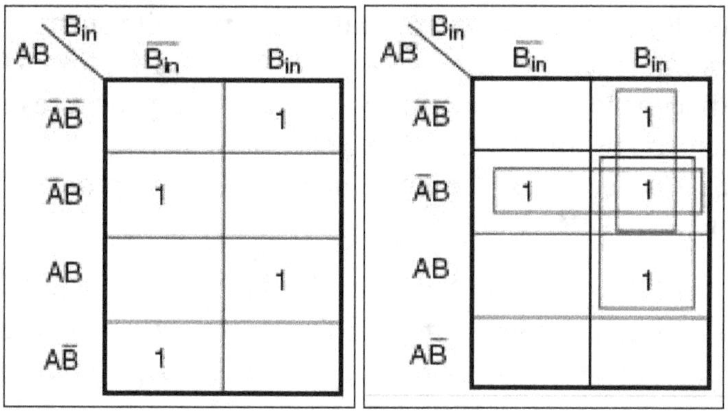

Full Subtractor Using Half Subtractors

The logic diagram for a full subtractor with two half subtractors and an OR gate is as follows.

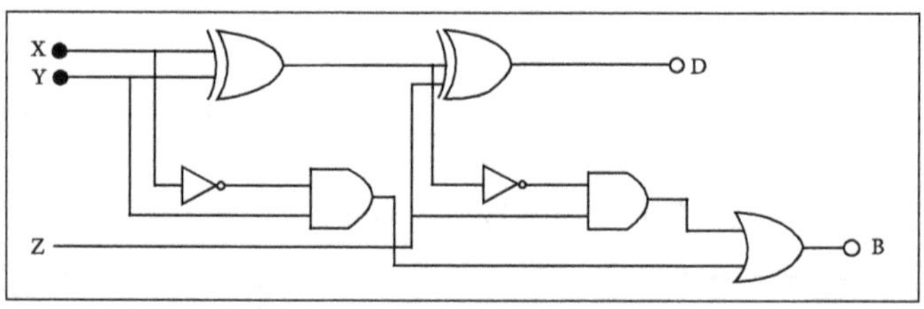

Full Subtractor Using Half Subtractor.

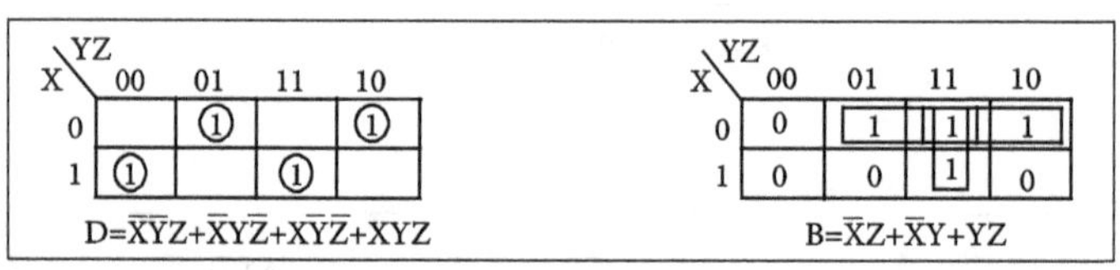

Serial Binary Adder

If speed is not of great importance, a cost-effective option is to use a serial adder. In Serial adder, bits are added a pair at a time (in one clock cycle).

$$A = a_{n-1}\ a_{n-2} \ldots\ldots\ a_{0}, B = b_{n-1}\ b_{n-2} \ldots\ldots\ldots\ b_{0}$$

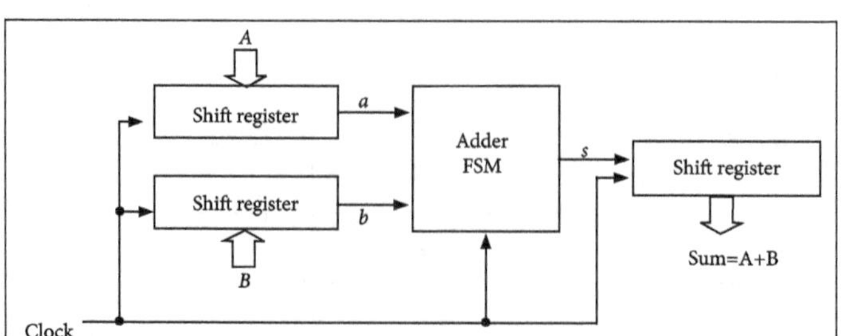

Block diagram for serial adder.

G: State that the carry-in is 0.

H: State that the carry-in is 1.

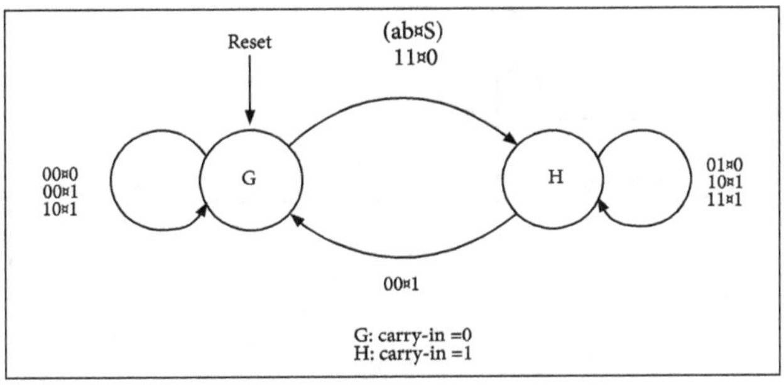

State diagram of the serial adder FSM.

State table:

Present state	Next state				Outputs			
	ab = 00 01 10 11				00 01 10 11			
G	G	G	G	H	0	1	1	0
H	G	H	H	H	1	0	0	1

$$Y = ab + ay + by$$

$$s = a \oplus b \oplus c$$

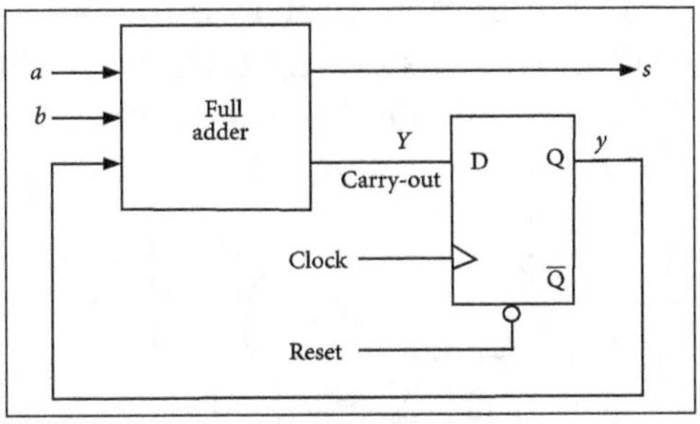

Circuit diagram.

The serial adder adds a pair of binary digits together with a carry bit from the previous addition. Each addition is triggered by a pulse on the clock input. The carry bit from the addition at the previous clock pulse can be set to zero with the carry reset pin 'R'. Some serial adders can optionally output the complement of the sum.

Parallel Binary Adder

Parallel adders are digital circuits that compute the addition of variable binary strings of equivalent or different sizes in parallel. The schematic diagram of a parallel adder is shown below in figure.

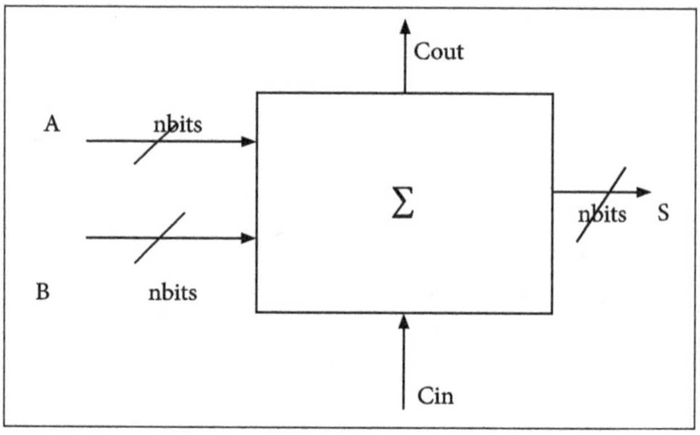

Parallel adder.

Parallel Binary Subtractor

To perform the subtraction of binary numbers with more than one bit is performed through the parallel subtractors. This parallel subtractor can be designed in several ways including combination of half and full subtractors, all full subtractors, all full adders with subtrahend complement input, etc. The below figure shows a 4 bit parallel binary subtractor formed by connecting one half subtractor and three full subtractors.

In this subtractor, 4 bit minuend $A_3A_2A_1A_0$ is subtracted by 4 bit subtrahend $B_3B_2B_1B_0$ and gives the difference output $D_3D_2D_1D_0$. The borrow output of each subtractor is connected as the borrow input to the next preceding subtractor. It is also possible to design a 4 bit parallel subtractor 4 full adders as shown in the below figure. This circuit performs the subtraction operation by considering that the addition of minuend and the complement of the subtrahend is equivalent to the subtraction process.

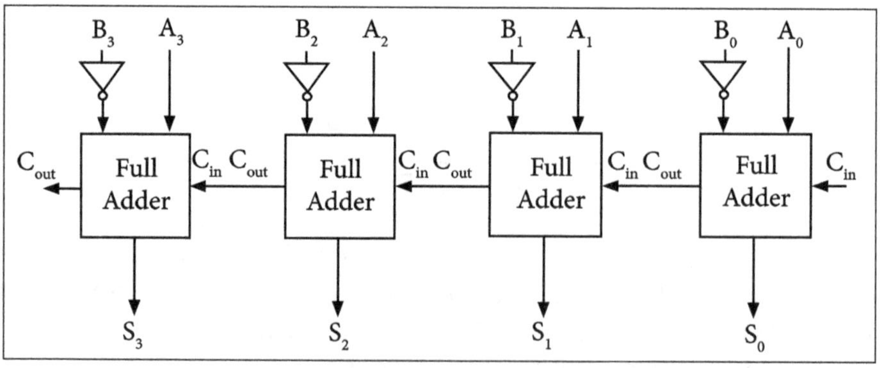

Full adders with subtrahend complement input.

We know that the subtraction of A by B is obtained by taking 2's complement of B and adding it to A. The two's complement of B is obtained by taking 1's complement and adding 1 to the least significant pair of bits. Hence in this circuit, 1's complement of B is obtained with the inverters (NOT gate) and 1 can be added to the sum through the input carry.

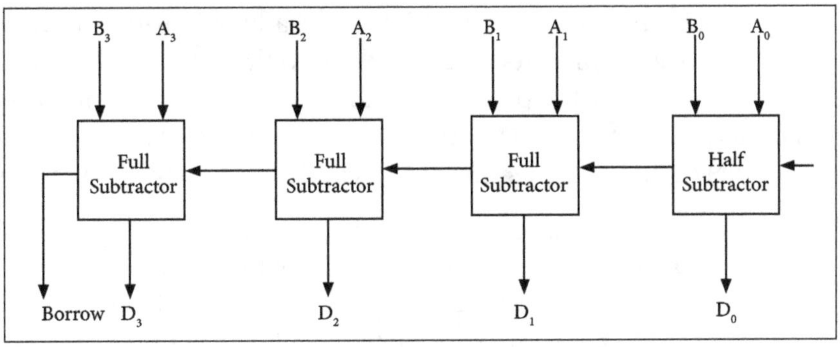

Full subtractors.

Multiplier

The combination circuits implemented to perform multiplication is called combinational multiplier. The advances in VLSI technology have made it possible to build combinational circuits that perform n x n bit multiplication for fairly large values of n.

Let us generalized the multiplication process for a 4 x 4 multiplier for two unsigned integers: multiplicand A = $A_3\ A_2\ A_1\ A_0$ and multiplier B = $B_3\ B_2\ B_1\ B_0$. As shown in Figure (a), each shifted multiplicand which is multiplied by either 0 or 1 depending on the corresponding multiplier bit is called partial product. Each partial product consists of four product component. The product is represented by a product term surrounded by rectangular box. The final 8-bit product is obtained by adding all partial products.

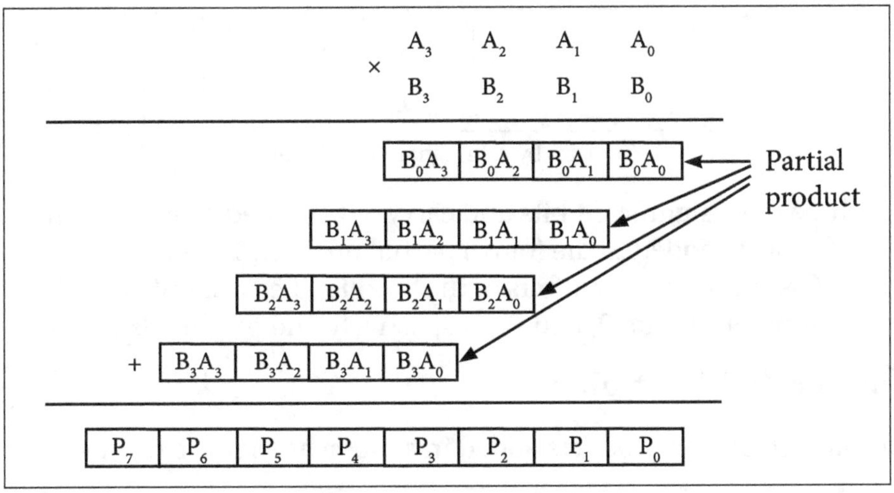

(a) 4 bit by 4-bit Binary Multiplication.

The first partial product is formed by multiplying B_0 by $A_3\ A_2\ A_1\ A_0$. The second partial product is formed by multiplying B_1 by $A_3\ A_2\ A_1\ A_0$. The third partial is formed by multiplying B_2 by $A_3\ A_2\ A_1\ A_0$ and the fourth partial product is formed by multiplying B_3 by $A_3\ A_2\ A_1\ A_0$. The multiplication of two bits such as B_0 and A_0 produces 1 if both bits are 1; otherwise, it produces 0. This is identical to AND operation. Therefore, the partial products can be implemented with AND gates as shown in Figure (b).

The 4-bit partial products are added using 4-bit parallel adder. During addition of first partial product, three most significant bits of it are added to the second partial product. As we take only three bits from first partial product, the fourth (most significant bit) is considered as 0. This is illustrated in Figure (b). The three most significant bits and carry out (treated as most significant bit) of first partial sum are then added to the third partial product.

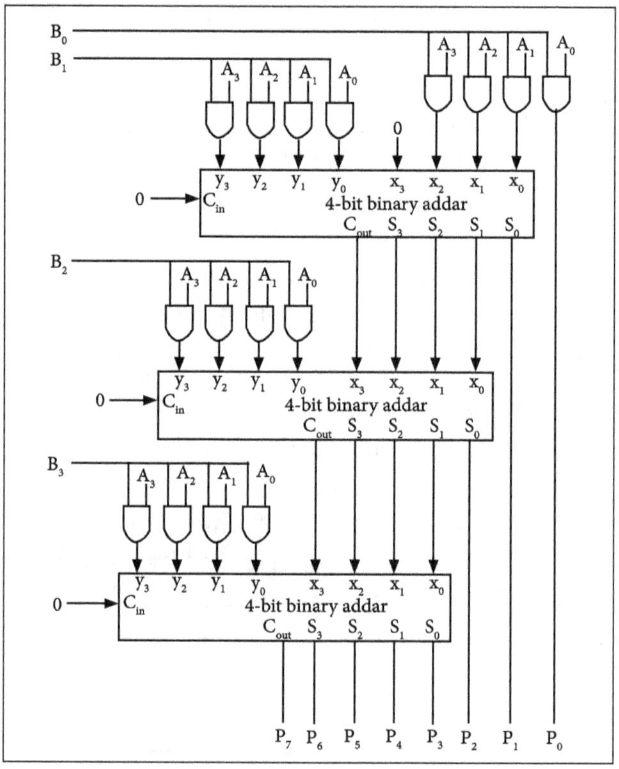

(b) 4-bit by 4-bit binary multiplier.

Finally, the three most significant bits and carry out (treated as most significant bit) of second partial sum are added to the fourth partial product. The carryout and third sum represents the five most significant bits of the product. Least significant bits of first and second partial sum represents P_1 and P_2, respectively and product B_0A_0 represents P_0.

Design of Parallel Multiplier

Multiplication with partial products added in as soon as they are formed.

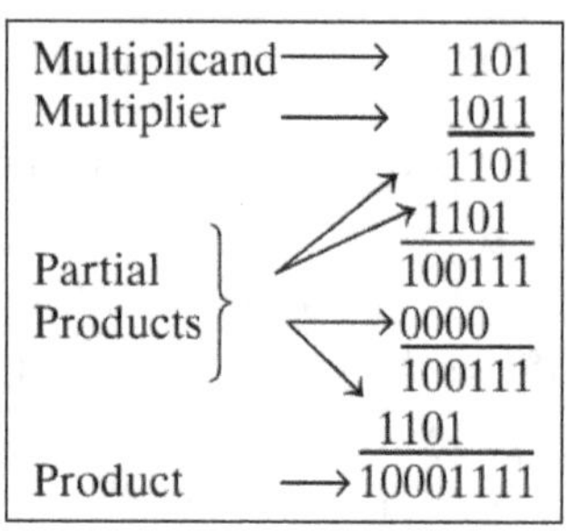

In every step:

- Multiplicand is shifted.
- Next bit of multiplier is examined.
- If this bit is 1, shifted multiplicand is added to the product.

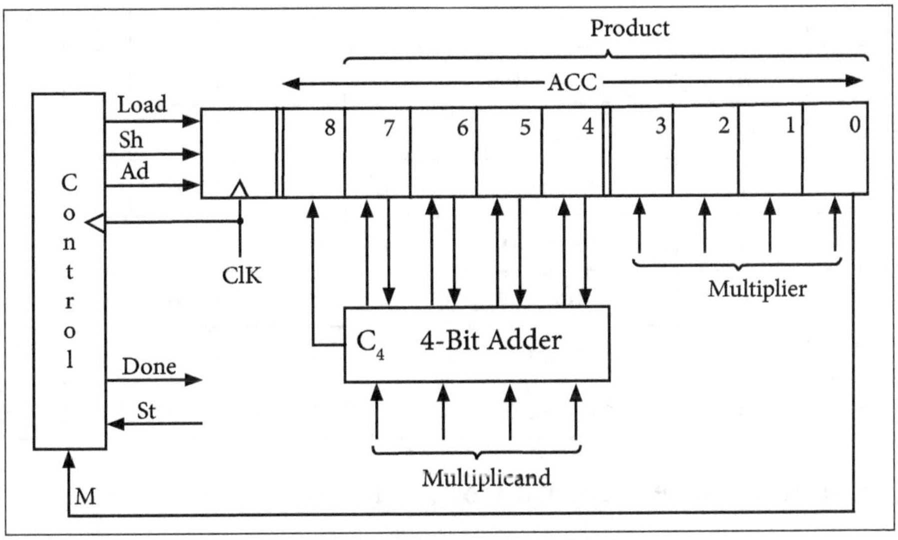

Block diagram of parallel multiplier.

Parallel Binary Multiplier Example:

initial contents of product register	0 0 0 0 0 ¦ 1 0 1 1 ←*M*
(add multiplicand because $M = 1$)	1 1 0 1
after addition	0 1 1 0 1 ¦ 1 0 1 1
after shift	0 0 1 1 0 1 ¦ 1 0 1 ←*M*
(add multiplicand because $M = 1$)	1 1 0 1
after addition	1 0 0 1 1 1 ¦ 1 0 1
after shift	0 1 0 0 1 1 1 ¦ 1 0 ←*M*
(skip addition because $M = 0$)	
after shift	0 0 1 0 0 1 1 1 ¦ 1 ←*M*
(add multiplicand because $M = 1$)	1 1 0 1
after addition	1 0 0 0 1 1 1 1 ¦ 1
after shift (final answer)	0 1 0 0 0 1 1 1 1 ¦
dividing line between product and multiplier →	

Operation of multiplier using counter:

Time	State	Counter	Product register	St	M	K	Load	Ad	Sh	Done
t_0	S_0	00	000000000	0	0	0	0	0	0	0
t_1	S_0	00	000000000	1	0	0	1	0	0	0
t_2	S_1	00	000001011	0	1	0	0	1	0	0
t_3	S_2	00	011011011	0	1	0	0	0	1	0
t_4	S_1	01	001101101	0	1	0	0	1	0	0
t_5	S_2	01	100111101	0	1	0	0	0	1	0
t_6	S_1	10	010011110	0	0	0	0	0	1	0
t_7	S_1	11	001001111	0	1	1	0	1	0	0
t_8	S_2	11	100011111	0	1	1	0	0	1	0
t_9	S_3	00	010001111	0	1	0	0	0	0	1

2.1.2 An Equality Detector and Comparator

An equality detector is a 8 bit magnitude comparator. This equality detector utilizes advanced silicon-gate CMOS technology to compare bit for bit two 8-bit words and indicate whether they are equal or not.

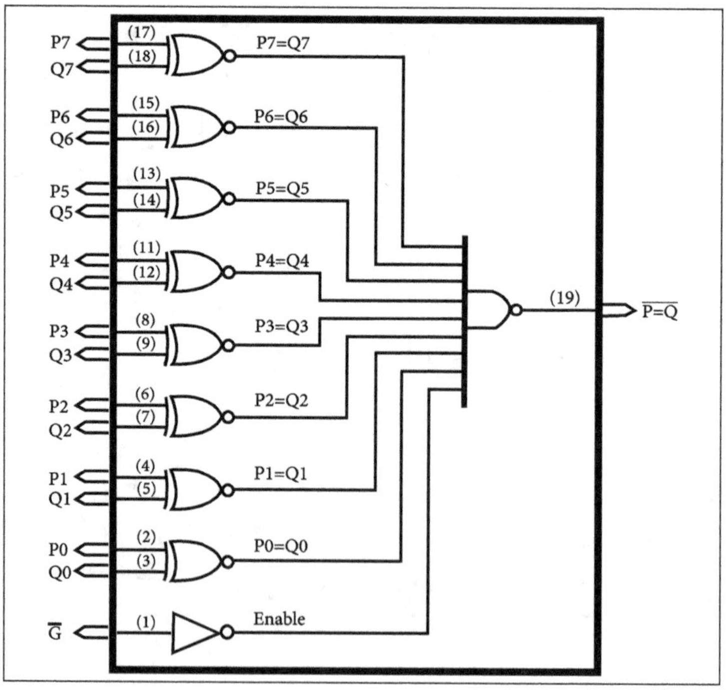

The $\overline{P=Q}$ output indicates the equality when it is low. A single active low enable is provided to facilitate the cascading of several packages and enable comparison of words greater than 8 bits. This device is useful in memory block decoding applications, where memory block enable signals must be generated from the computer address information.

The comparator combines the low power consumption of CMOS, but the inputs are compatible with TTL logic levels, and the output can drive 10 low power Schottky equivalent loads. These devices are intended to interface between TTL and NMOS components and standard CMOS devices. These parts are also plug in the replacements for LS-TTL devices and can be used to reduce the power consumption in the existing designs. All the inputs are protected from damage due to static discharge by diodes to VCC and ground.

Features

- Typical propagation delay: 20 ns.
- TTL input compatible.
- Large output current: 4 mA.
- Low quiescent current: 80 μA maximum (74HCT Series).
- Same as HCT521.

Truth table:

Inputs		Output
Data	Enable	$\overline{P=Q}$
P, Q	$\overline{Q}$	
P = Q	L	L
P > Q	L	H
P < Q	L	H
X	H	H

Comparator

It is a combinational circuit that compares two numbers and determines their relative magnitude. The output of comparator is usually 3 binary variables indicating:

A > B

A = B

A < B

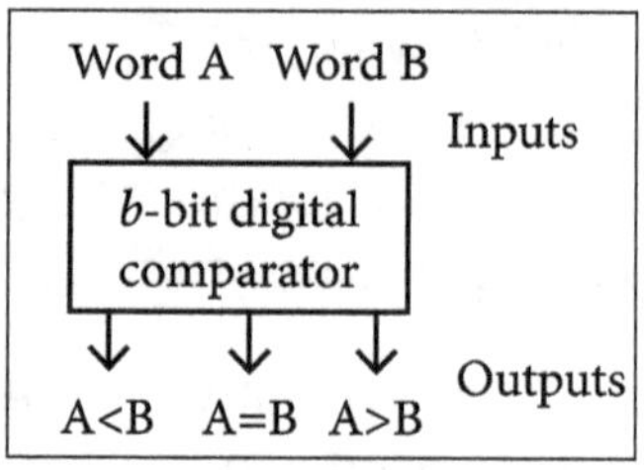

Comparator

Digital comparator is a combinational circuit, designed to compare the two n-bit binary words applied at its input. Comparator has three outputs A > B, A = B and A < B depending on the result of comparison of one of their outputs.

4-Bit Digital Comparator

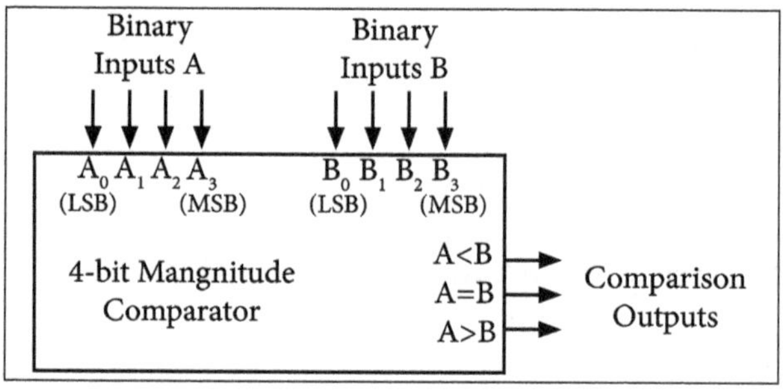

4-Bit Digital Comparator.

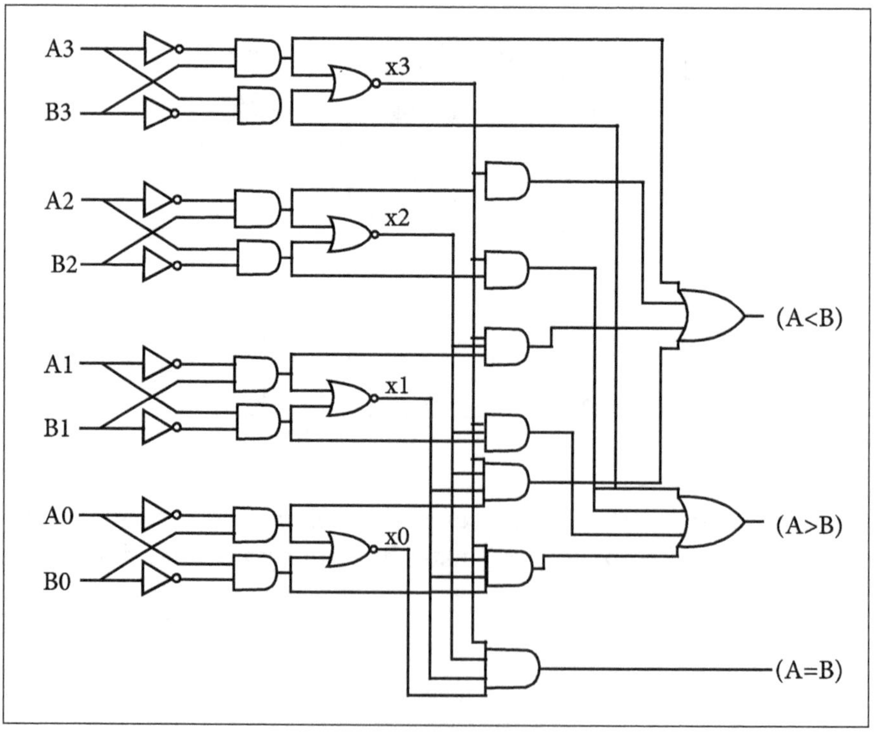

4-bit magnitude comparator.

The method for binary numbers with more than 2 bits can also be found in the similar way. The above figure shows the 4-bit magnitude comparator.

Input,

$A = A_3A_2A_1A_0$

$B = B_3B_2B_1B_0$

From the above diagram the input are as follows,

A = B: A3=B3, A2=B2, A1=B1, Ao=Bo

xi = AiBi + Ai'Bi'

XOR-Invert = (AiBi' + Ai'Bi)' = (Ai' + Bi) (Ai + Bi')

= Ai'Ai + Ai'Bi' + AiBi + BiBi'

= AiBi + Ai'Bi'

Output: x3x2x1xo,

A> B

Output: A3B'3 + x3A2B'2 + x3x2A1B'1 + x3x2x1AoB'o,

(A< B)

Output: A'3B3 + x3A'2B2 + x3x2A'1B1 + x3x2x1A'oBo

Truth table:

Comparing Inputs				**Outputs**		
A3,B3	A2,B2	A1,B1	Ao,Bo	A > B	A < B	A = B
A3 > B3	X	X	X	H	L	L
A3 < B3	X	X	X	L	H	L
A3 = B3	A2 > B2	X	X	H	L	L
A3 = B3	A2 < B2	X	X	L	H	L
A3 = B3	A2 = B2	A1 > B1	X	H	L	L
A3 = B3	A2 = B2	A1 < B1	X	L	H	L
A3 = B3	A2 = B2	A1 = B1	Ao > Bo	H	L	L
A3 = B3	A2 = B2	A1 = B1	Ao < Bo	L	H	L
A3 = B3	A2 = B2	A1 = B1	Ao = Bo	H	L	L
A3 = B3	A2 = B2	A1 = B1	Ao = Bo	L	H	L
A3 = B3	A2 = B2	A1 = B1	Ao = Bo	L	L	H
H = High Voltage Level; L = Low Voltage Level; X = Don't Care						

2.1.3 Line Decoder

Decoder

A decoder is a combinational circuit that converts binary information from n input lines to a maximum of 2n output lines. A decoder is a multiple input multiple output logic circuit that converts coded inputs into coded outputs where the input and output codes are different.

n-to-2n Line Decoder

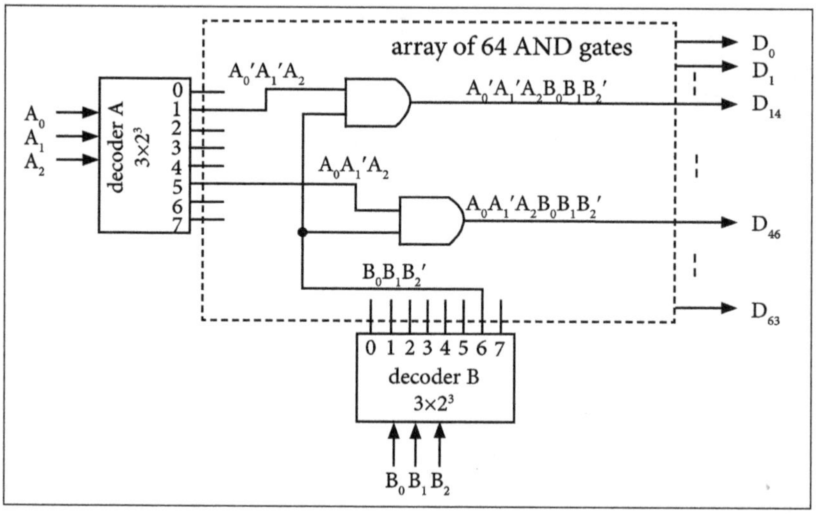

n-to-2n-Line Decoder.

Each AND gate now has only n (in this case three) inputs and there are 2n (in this case eight) outputs. Since there is no data input line to control, what used to be control inputs no longer serve that function. Instead, they are the data inputs to be decoded. This circuit is an example of an n-to-2n line decoder. Each output represents the minterm.

Output k is 1 whenever the combination of the input variable values is the binary equivalent of decimal k. Now, suppose that the data input line from the demultiplexer is not removed but retained and viewed as an enable input, the decoder now operates only when the enable is 1.

Viewed conversely, an n-to-2n-line decoder with an enabled input can also be used as a demultiplexer, where the enable becomes the serial data input and the data inputs of the decoder become the control inputs of the demultiplexer.

Decoders are available as integrated circuits (MSI) whose n = 3 and n = 4 are quite common. However, there will be always some practical limitations on the fan-in decoders of higher order and are often designed using lower-order decoders interconnected with a network of other gates.

3 To 8 Line Decoder

3 inputs are decoded into eight outputs. Each output represents one of the minterms of the 3 input variable.

EN	A	B	C	Y_7	Y_6	Y_5	Y_4	Y_3	Y_2	Y_1	Y_0
0	X	X	X	0	0	0	0	0	0	0	0
1	0	0	0	0	0	0	0	0	0	0	1
1	0	0	1	0	0	0	0	0	0	1	0
1	0	1	0	0	0	0	0	0	1	0	0
1	0	1	1	0	0	0	0	1	0	0	0
1	1	0	0	0	0	0	1	0	0	0	0
1	1	0	0	0	0	0	1	0	0	0	0
1	1	0	1	0	0	1	0	0	0	0	0
1	1	1	0	0	1	0	0	0	0	0	0
1	1	1	1	1	0	0	0	0	0	0	0

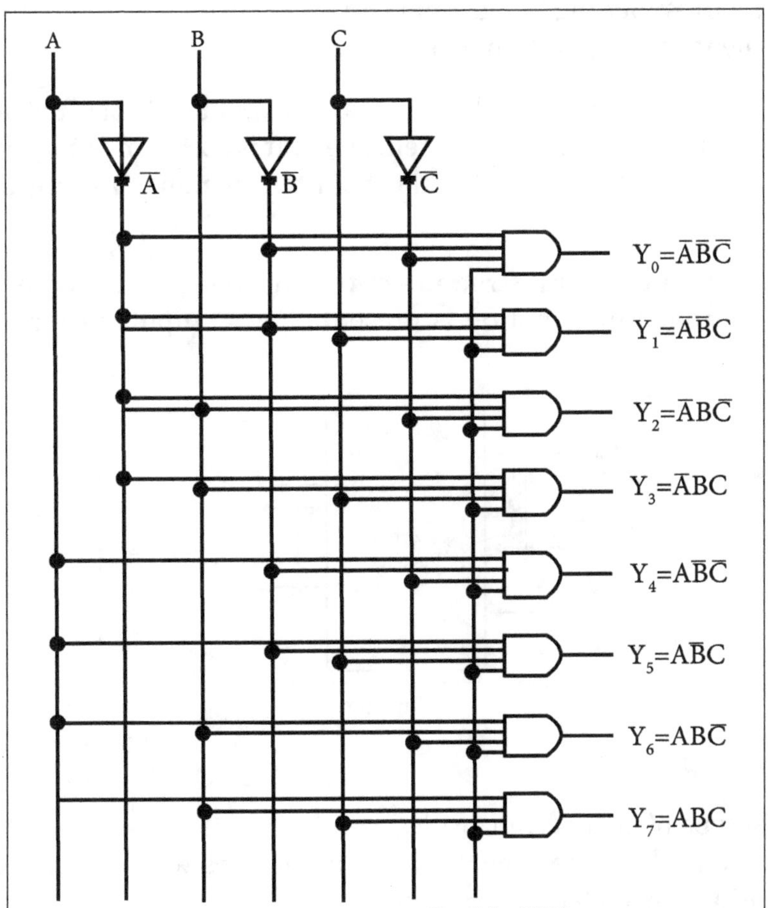

3 to 8 line decoder.

Difference between Decoder and Demultiplexer

<table>
<tr><th>Decoder</th><th>Demultiplexer</th></tr>
<tr><td>(i) Schematic diagram
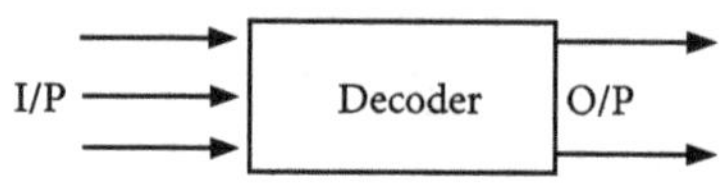
</td><td>Schematic diagram
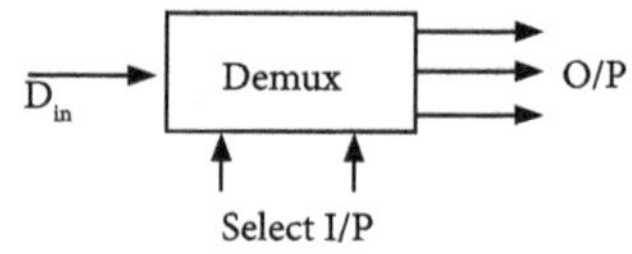
</td></tr>
<tr><td>(ii) Application: Used as code converter.</td><td>Used as a distributor switch to implement Boolean expressions.</td></tr>
<tr><td>(iii) There are no selection lines.</td><td>The selection is controlled by value of selection.</td></tr>
</table>

2.1.5 Encoders

An encoder is the digital circuit that performs the inverse operation of a decoder. An encoder has 2^n input lines and n output lines. In the encoder, the output lines generate the binary code corresponding to the input value. An encoder is a combinational circuit that performs the inverse operation of a decoder.

When the device output code has fewer bits than the input code has, the device is usually called an encoder. e.g., 2^n-to-n priority encoders. It is a 2^n-to-n binary encoder, where it has only 2n inputs and the output is the n-bit binary number corresponding to the active input.

Figure below shows the general structure of the encoder circuit. As shown in the figure below, the decoded information is presented as 2^n inputs producing n possible outputs.

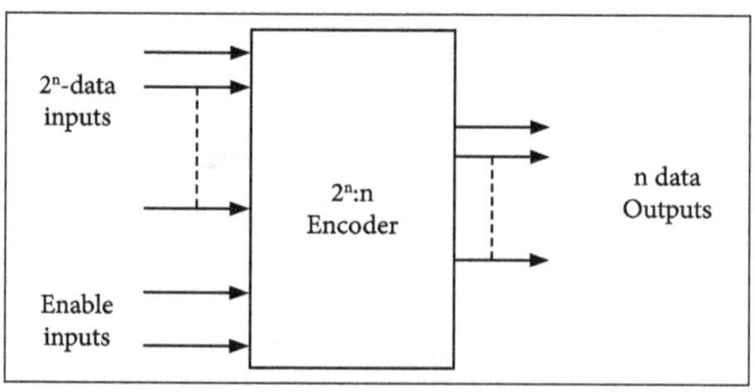

Encoder

Priority Encoder

This is a special type of encoder. Priorities are given to the input lines. If two or more inputs lines are '1' at the same time, then the input lines with highest priority will be considered. A priority encoder is the practical form of an encoder. The encoders that is available in IC form are all priority encoders.

In this type of encoder, a priority is assigned to each of the input so that when more than one input is simultaneously active, the input with the highest priority is encoded. Let us also assume that the input lines D_2, D_4 and D_7 are all simultaneously in logic 1' state. In such cases, only D_7 will be encoded and its output will be 111.

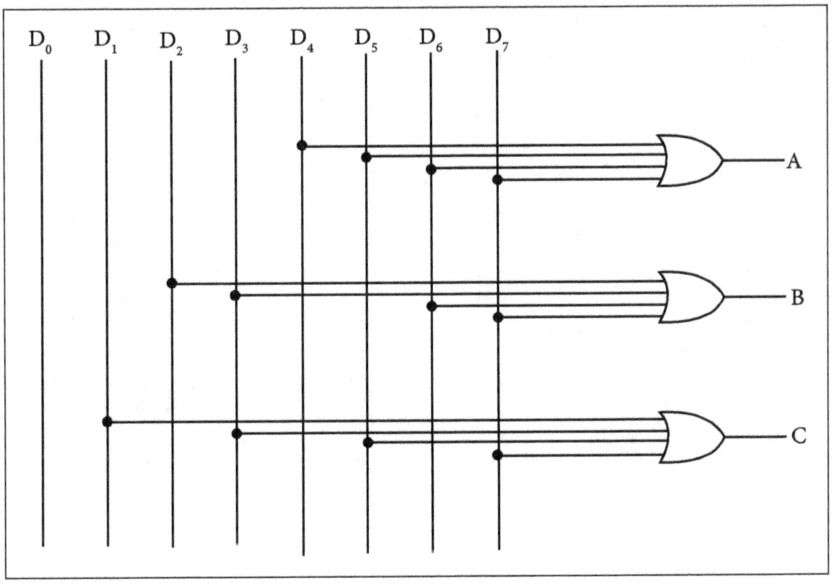

Priority Encoder.

Decimal to BCD Encoder

The decimal to BCD encoder, usually has ten input lines and four output lines. The decoded decimal data acts as an input for encoder and encoded BCD output is available on the four output lines.

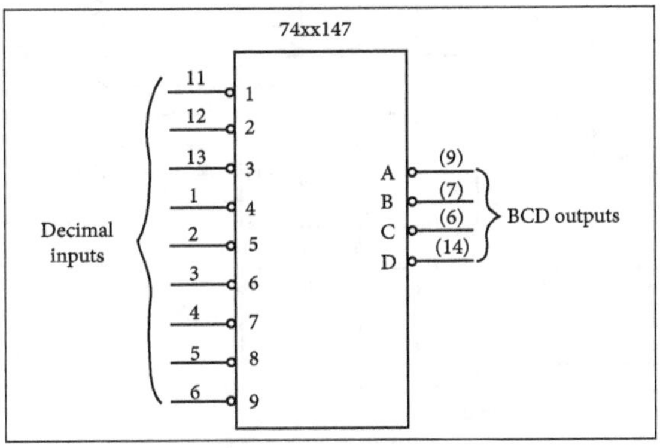

Logic symbol for 74xx147 (Decimal to BCD encoder).

The above figure shows the logic symbol for decimal to BCD encoder IC, IC 74xx147. It has nine input lines and four output lines. Both input and output lines are asserted active low. It is important to note that there is no input line for decimal zero. When this condition occurs, all output lines are 1. The function table for the 74xx147 is shown in the below table.

Truth table for decimal to BCD encoder:

Decimal value	Input									Output			
	1	2	3	4	5	6	7	8	9	D	C	B	A
0	1	1	1	1	1	1	1	1	1	1	1	1	1
1	0	1	1	1	1	1	1	1	1	1	1	1	0
2	X	0	1	1	1	1	1	1	1	1	1	0	1
3	X	X	0	1	1	1	1	1	1	1	1	0	0
4	X	X	X	0	1	1	1	1	1	1	0	1	1
5	X	X	X	X	0	1	1	1	1	1	0	1	0
6	X	X	X	X	X	0	1	1	1	1	0	0	1
7	X	X	X	X	X	X	0	1	1	1	0	0	0
8	X	X	X	X	X	X	X	0	1	0	1	1	1
9	X	X	X	X	X	X	X	X	0	0	1	1	0

X Denotes Don't Care Condition

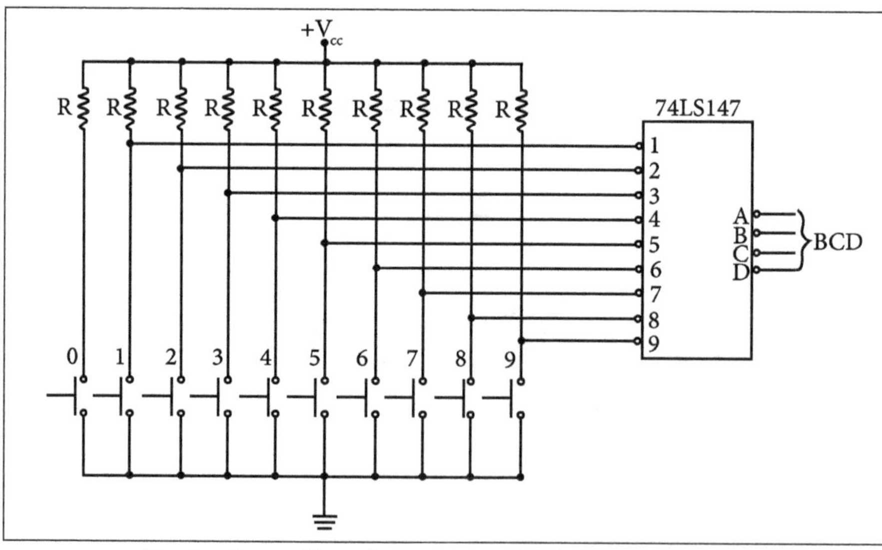

Ten key keypad interface using decimal to BCD encoder.

2.1.6 Multiplexers and De-multiplexers

Multiplexer

Multiplexer is a digital switch. It allows digital information from several sources to be routed onto a single output line. Multiplexer is a special type of combinational circuit. It has n data inputs, one output and select inputs with $2^m = n$.

Depending on the digital code applied at the select inputs, one out of a data sources is selected and transmitted to the single output Y. E is called as a strobe or enable input which is useful for cascading. It is generally an active low terminal that means it will perform required operation when it is low.

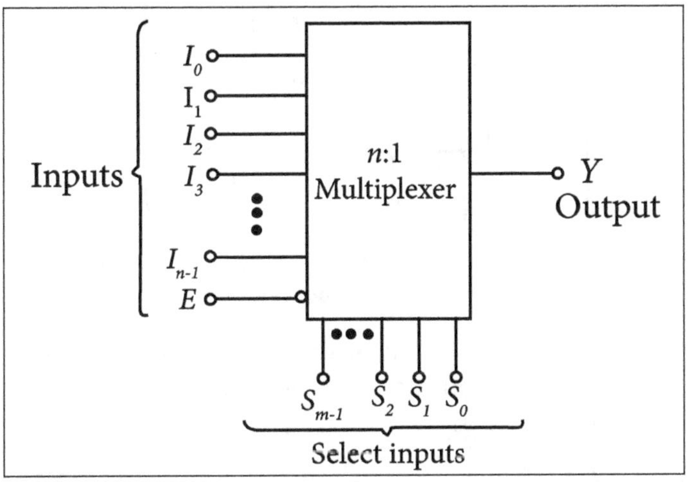

Multiplexer

Applications of Multiplexer

It is used:

- In the implementation of combinational circuits.
- As function generators.

Data Routing

Multiplexers can be used to route the data from one of several sources to one destination.

Control Sequences

A Multiplexer can also be used as a part of control sequences.

2 X 2 Multiplexer

Multiplication of binary number is performed in the same way as in decimal numbers. The multiplicand is multiplied by each bit of the multiplier which starts from the least significant bit. Each such multiplication forms a partial product. Successive partial products are shifted one position to the left. The final product is obtained from the sum of partial products.

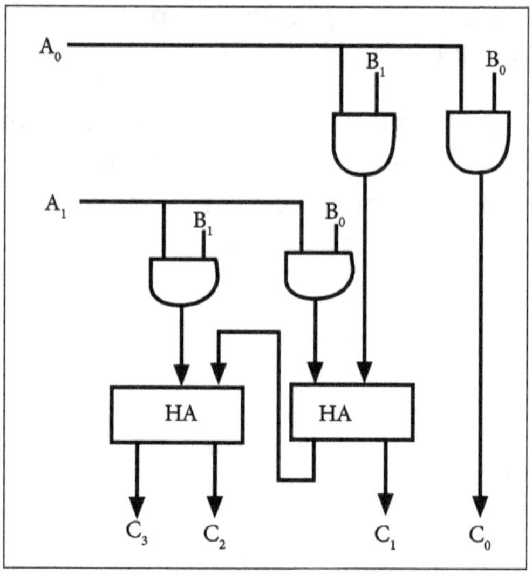

2 x 2 multiplexer.

The multiplication of 2 bit number is considered. The multiplicand bits are B_1 and B_0, the multiplier bits are A_1 and A_0 and the product is C_3 C_2 C_1 C_0. The first partial product is formed by multiplying Ao by B_1B_0.

The multiplication of two bits such as A_0 and B_0 produces 1 if both bits are 1, otherwise produces 0. This is identical to AND Operation. Therefore, the partial product can be implemented with AND gates as,

$$\begin{array}{cccc} & & B_1 & B_0 \\ & & A_1 & A_0 \\ \hline & & A_0B_1 & A_0B_0 \\ & A_1B_1 & A_1B_0 & \\ \hline C_3 & C_2 & C_1 & C_0 \end{array}$$

The second partial product is formed by multiplying A1 by B1Bo and shifted one position to left. The two partial products are added with two half adder circuits. The LSB of the product does not have to go through an adder since it is formed by the output of the first AND gate.

Demultiplexers

In addition to the data input there must be other inputs to control the transmission of the data to the appropriate data output line at any given time. This demultiplexer circuit having eight output lines is shown in Figure (a). For the same number of control inputs, it has same number of AND gates. But now each AND gate output is a circuit output.

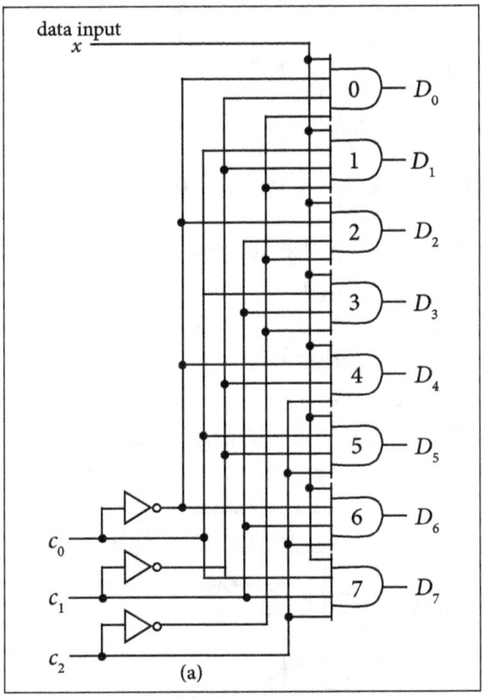

(a) Demux circuit.

Relatively, each gate having its own separate data input, single data line will form one of the inputs to each AND gate, the other AND inputs being control inputs. When the word formed by the control inputs $C_2C_1C_0$ which is the binary equivalent of decimal k, then the data input x is routed to output Dk. Otherwise, for a demultiplexer which has n control inputs, each AND gate output shall correspond to a minterm of n variables.

For a given combination of control inputs, only one minterm can take on the value 1, the data input is routed to the AND gate corresponding to this minterm. For example, if the logical expression for the output D_3 is $xC_2'C_1C_0$. Thus, when $C_2C_1C_0$ = 011, then D_3 = x and all other D_i are equal to 0.

The complete truth table for the eight-output demultiplexer is given below:

Truth table:

Control inputs			Data outputs							
C_2	C_1	C_0	D_0	D_1	D_2	D_3	D_4	D_5	D_6	D_7
0	0	0	x	0	0	0	0	0	0	0
0	0	1	0	x	0	0	0	0	0	0
0	1	0	0	0	x	0	0	0	0	0
0	1	1	0	0	0	x	0	0	0	0
1	0	0	0	0	0	0	x	0	0	0

1	0	1	0	0	0	0	0	x	0	0
1	1	0	0	0	0	0	0	0	x	0
1	1	1	0	0	0	0	0	0	0	x

Higher Order Demultiplexers and Multiplexers

Since 4-line-to 16-line decoders are the largest available circuits in ICs, to meet the larger inputs need there should be a provision for expansion. This is made possible by using enable input terminal. Figure (a) shows a 5-line-to 32-line decoder and figure (b) shows a 8-line-to 256-line decoder using 4-line-to 16-line decoders. Similarly, an m-line-to-n-line decoder can be implemented. However, if only a few codes of a large number need be recognised, the alternative approach, such as the one shown in Figure (c), can be used. This is connected to detect the digital number 00011111.

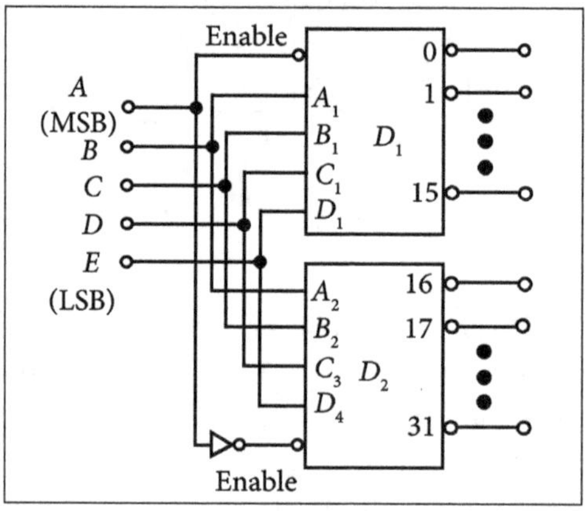

(a) 5-line-to 32-line decoder using two 4-line-to 16-line decoders.

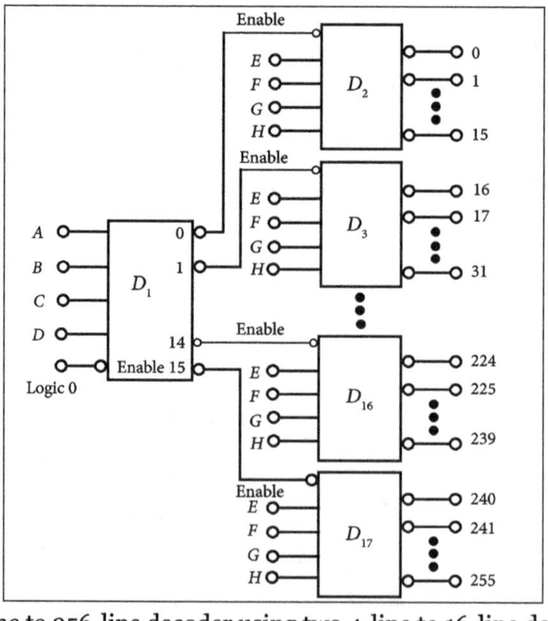

(b) 8-line to 256-line decoder using two 4-line to 16-line decoders.

The most-significant 4-bits are applied at ABCD inputs and the least-significant bits are applied at EFGH inputs. The output goes low when the most-significant bits are 0001 and the least-significant bits are 1111. The circuit can be expanded to detect other 8-bit codes.

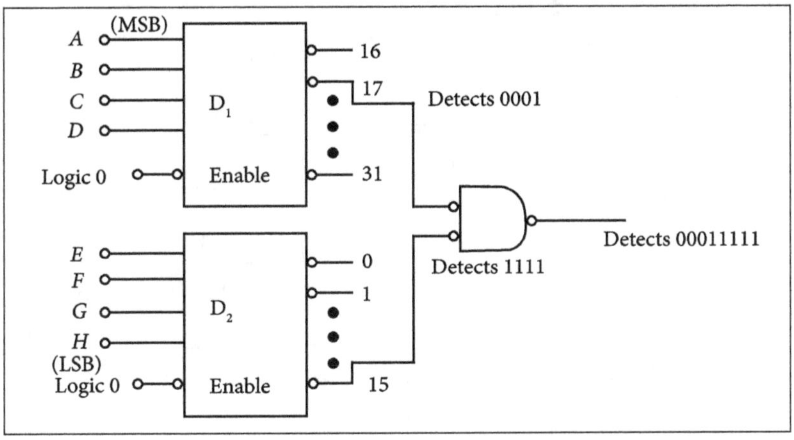

(c) An alternative approach to decode combinations.

Higher Order Multiplexing

Thc 16-to-1 multiplexers are the largest available ICs. Hence, to meet the larger input needs there must be a provision for expansion. This can be achieved with the help of enable or strobe inputs and multiplexer stacks or trees are designed. Two commonly used methods for this purpose are illustrated in Figure (a) and (b).

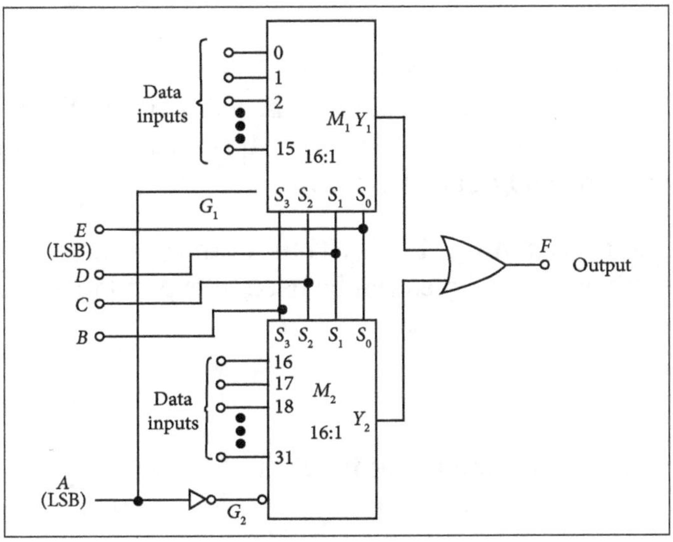

(a) 16 : 1 mux with OR gate.

The circuit of Figure (a) uses two 16:1 multiplexers (M_1 and M_2) for the realization of a 32:1 multiplexer. The lower order of 16 data input lines (I_0 - I_{15}) are applied at the data input terminals of the multiplexer M1 and the higher order 16 data input lines (I_{15} - I_{31}) are applied at the data input terminals of the multiplexer M_2.

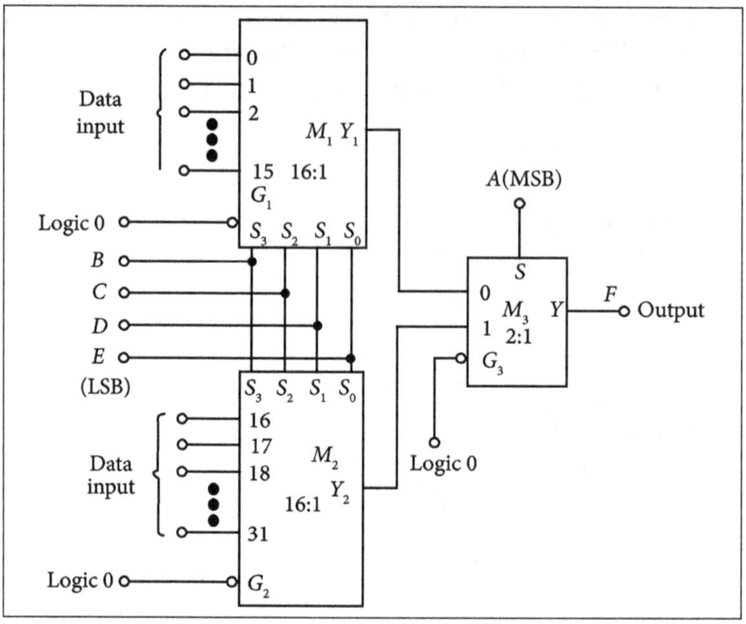

(b) 16 : 1 mux with 2 :1 mux.

For a 32:1 multiplexer, the number of select input lines is required to be 5. The 5-bit select inputs are ABCDE. The most-significant select input bit A is applied at G1 and $\bar{A}$ is applied at G_2. B, C, D and E are connected to S_3, S_2, S_1 and So inputs of both the multiplexers. The output F of the multiplexer is obtained by using an OR gate, where $F = Y_1 + Y_2$.

When A = 0, the multiplexer M_1 is enabled and M_2 is disabled, thereby allowing one of the lower order 16 bits to Y_1, depending upon the value of BCDE. Y_2 = 0. Therefore, F = Y_1. Similarly, when A = 1, M2 is enabled and M_1 is disabled, then F = Y_2. Figure (b) uses a 2:1 multiplexer instead of an OR gate.

When A = 0, F = Y_1 and when A = 1, F = Y_2. Thus, both the circuits perform similar operation. These two general techniques can be used to expand to an n input multiplexer without any difficulty.

Problems

1. Here let us implement the following function $f(A,B,C) = \sum m(0,3,5)$ using 4: 1 multiplexer.

Solution:

Given:

$$f(A, B, C) = \sum m(0, 3, 5)$$

Formula to be used:

Input lines = 2^{n-1}

Variables n = 3 (A, B, C)

Select lines = n – 1 = 2(S_0, S_1)

2^{n-1} to 1 MUX= 4: 1 MUX

Input lines = 2^{n-1} = 4(I_0, I_1 I_2, I_3)

Implementation Table:

	I_0	I_1	I_2	I_3
$\bar{A}$		1	2	
A	4		6	7
	$\bar{A}$	A	0	$\bar{A}$

$I_0 = \bar{A}, I_1 = A, I_2 = 0, I_3 = \bar{A}$

Multiplexer implementation

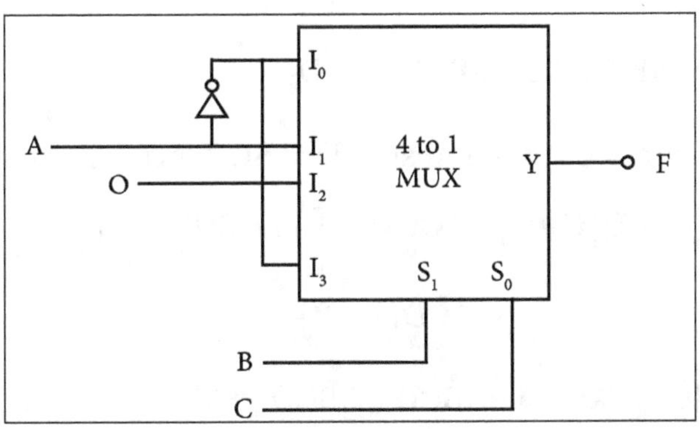

2. Let us implement $F(x, y, z) = \sum(1,2,6,7)$ using 4 × 1 MUX.

Solution:

Given:

	D_0	D_1	D_2	D_3
$\bar{x}$	0	①	②	3
x	4	5	⑥	⑦
	0	$\bar{x}$	1	x

$$F(x,y,z) = \sum(1, 2, 6, 7)$$

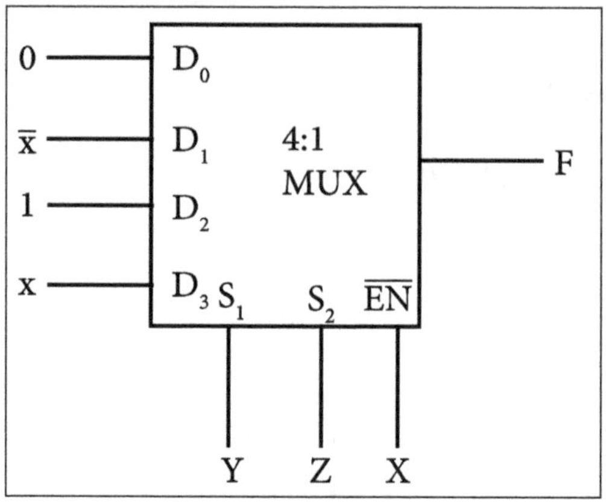

4:1 MUX.

3. Let us see the implementation of Boolean function using 8:1 multiplexer

$$F(A,B,C,D) = \bar{A}B\bar{D} + ACD + \bar{B}CD + \bar{A}\bar{C}D$$

Solution:

Given:

$$F(A,B,C,D) = \bar{A}B\bar{D} + ACD + \bar{B}CD + \bar{A}\bar{C}D$$

The given Boolean expression is not in standard SOP form. The conversion to SOP is,

$$F(A,B,C,D) = \bar{A}B\bar{D}(C+\bar{C}) + ACD(B+\bar{B}) + \bar{B}CD(A+\bar{A})$$

$$+\bar{A}\bar{C}D(B+\bar{B})$$

$$= \bar{A}BC\bar{D} + \bar{A}B\bar{C}\bar{D} + ABCD + A\bar{B}CD + A\bar{B}CD$$

$$+\bar{A}\bar{B}CD + \bar{A}B\bar{C}D + \bar{A}\bar{B}\bar{C}D$$

$$= \bar{A}BC\bar{D} + \bar{A}B\bar{C}\bar{D} + ABCD + A\bar{B}CD$$

$$+\bar{A}\bar{B}CD + \bar{A}B\bar{C}D + \bar{A}\bar{B}\bar{C}D$$

The Truth Table for this standard SOP is:

No.	Minterms	A	B	C	D	Y
0		0	0	0	0	0
1	$\bar{A}\bar{B}\bar{C}D$	0	0	0	1	1

2		0	0	1	0	0
3	$\bar{A}\bar{B}CD$	0	0	1	1	1
4	$\bar{A}B\bar{C}\bar{D}$	0	1	0	0	1
5	$\bar{A}B\bar{C}D$	0	1	0	1	1
6	$\bar{A}BC\bar{D}$	0	1	1	0	1
7		0	1	1	1	0
8		1	0	0	0	0
9		1	0	0	1	0
10		1	0	1	0	0
11	$A\bar{B}CD$	1	0	1	1	1
12		1	1	0	0	0
13	-	1	1	0	1	0
14		1	1	1	0	0
15	ABCD	1	1	1	1	1

Implementation Table:

	D_0	D_1	D_2	D_3	D_4	d_5	d_6	d_7
$\bar{A}$	0	1	2	3	4	5	6	7
A	8	9	10	11	12	13	14	15
Y	0	$\bar{A}$	0	1	$\bar{A}$	$\bar{A}$	$\bar{A}$	A

MUX Implementation

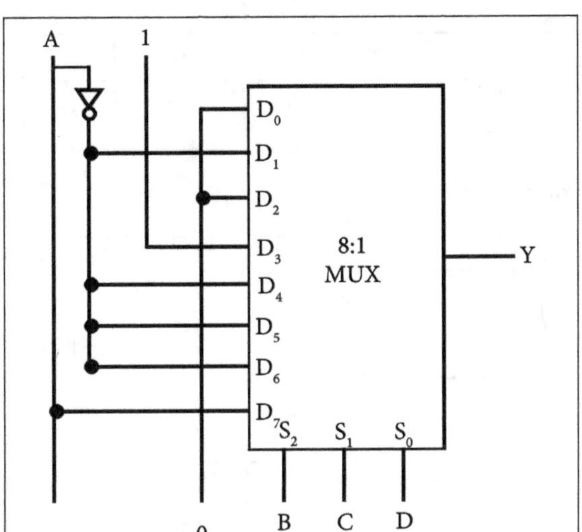

4. Let us implement the following with a multiplexer,

$$F(A, B, C) = \sum(1, 2, 4, 5)$$

Solution:

Given:

$$F(A, B, C) = \sum(1, 2, 4, 5)$$

Two of the variable B and C, are applied to the selection lines S1 and So respectively.

Truth table:

Minterm	A	B	C	F
0	0	0	0	0
1	0	0	1	1
2	0	1	0	1
3	0	1	1	0
4	1	0	0	1
5	1	0	1	1
6	1	1	0	0
7	1	1	1	0

Implementation Table:

	D_0	D_1	D_2	D_3
$\overline{A}$	0	①	②	3
A	④	⑤	6	7
	A	1	$\overline{A}$	0

The multiplexer implementation is given as,

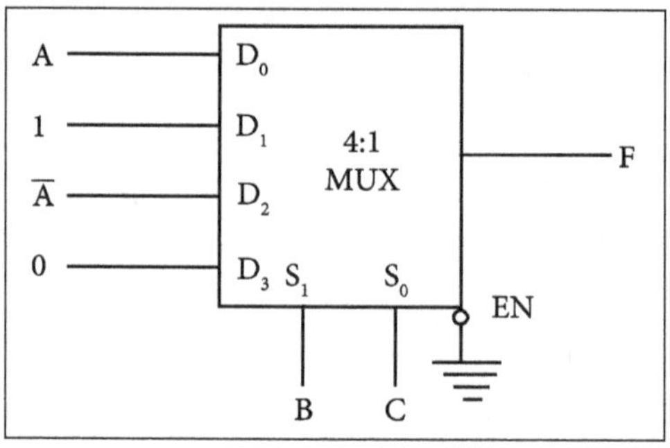

2.2 Synchronous Sequential Logic Design

Combinational circuits and systems create an output based on input variables only. Sequential circuits use current input variables and previous input variables by storing the information and putting back into the circuit on the next clock (activation) cycle.

There are 2 types of input to the combinational logic:

External Inputs

The inputs which come from outside the circuit design and are not controlled by the circuit.

Internal Inputs

The inputs which are a function of a previous output states. The internal inputs and outputs are referred as "secondaries". Secondary inputs are the state variables produced by the storage elements, where as secondary outputs are excitations for the storage elements.

Classifications of Sequential Circuits

The sequential circuits are classified on the basis of timing of their signals into two types. They are:

- Synchronous sequential circuit.
- Asynchronous sequential circuit.

Synchronous types use a pulsed or level inputs and a clock input to drive the circuit (with restrictions on pulse width and circuit propagation).

Asynchronous sequential circuits do not use a clock signal as synchronous circuits. Instead the circuit is driven by the pulses of the inputs.

A pulsed output is an output that lasts for the duration of a particular input pulse but can be less in some cases. For the clocked sequential circuits, the output pulse is the same duration as the clock pulse. A level output refers to an output that changes the state at the start of an input pulse or clock pulse and remains in that state until the next input or clock pulse.

Synchronous Sequential Circuit

A synchronous sequential circuit is a system whose behavior can be defined from the knowledge of its signals at discrete instants of time.

Present State and Next State

The information stored in memory elements at any given time defines the present state of sequential circuit. The present state and the external inputs determine the outputs and next state of the sequential circuits.

Secondary Variables

The delay elements provide a short term memory for the sequential circuit. The present state and. next state variables in asynchronous sequential circuits are called secondary variables.

Comparison of combinational and sequential circuit:

Combinational Circuits	Sequential Circuits
The output depends only on the input present at that input.	The output depends not only on the inputs but also on the past outputs.
There is no feedback.	Output is fed back to input.
Memory unit is not required.	Memory unit is required.
Eg: MUX, Demux, Encoder, Decoder.	Eg: Shift registers, counters.

Analysis

It consists of a table or a diagram for the time sequence of inputs, outputs and interval states. A logic diagram is recognized as a synchronous sequential circuit if it includes flip - flops. The flip - flops may be of any type and the logic diagram may (or) may not include combinational circuit gates.

Design

It consists of choosing the flip flops and combinational gate structure together with the flip - flops. The number of flip - flop is determined from the number of states needed in the circuit. The combinational circuit is derived from the state table.

2.3 Storage Elements

Sequential Circuits are built out of combinational logic and one or more memory or storage elements.

Examples: Memories, Registers, Counters and Control Unit.

Combinational logic cannot remember. Its output logic values are function of inputs only. Thus, feedback is needed to remember a logic value. Memory elements are needed in most digital logic circuits to hold logic values. Two basic types of memory elements are as follows:

- Latches: Level-sensitive to inputs.
- Flip-flops: Edge-triggered on active edge of clock.

Latches

The Latches and the flip-flops are the basic elements for storing information. One latch or flip-flop can store 1 bit of information. The major difference between latches and flip-flops is that for latches, their outputs are continuously affected by their inputs as long as the enable signal is asserted.

In other words, when they are enabled, their content modifies immediately when their inputs change. Flip-flops have their content change only either at the rising or falling edge of the enable signal. This enable signal is generally the controlling clock signal. After increasing or decreasing the edge of the clock, the flip-flop content remains constant even if the input changes.

There are basically 4 main types of latches and flip-flops such as SR, D, JK and T. The important differences in these flip-flop types are the number of inputs they have and how they change state. For each type, there are also different variations that enhance their operations.

SR Latch

The bistable elements are able to remember or store one bit of information. However, because it does not have any inputs, we can't change the information bit that is stored in it. In order to change the information bit, we require to add inputs to the circuit. The simplest way to add inputs is to replace the two inverters with 2 NAND gates as represented in Figure (a). This circuit is called a SR latch. In addition to the 2 outputs Q and Q', there are two inputs S' and R' for set and reset respectively. Following the convention, the prime in S and R denotes that these inputs are active low. The SR latch can be in one of 2 states: a set state when Q = 1 or a reset state when Q = 0.

To make the SR latch go to the set state, we simply assert the S' input by setting it to 0. Remember that 0 NAND provides a1, hence Q = 1 and the latch is set. If R' is not asserted (R' = 1), then the output of the bottom NAND gate will provide a0 and so Q' = 0. This situation is represented in Figure (d) at time t0.

If we de-assert S' so that S' = R' = 1, the latch will remain at the set state because Q', the second input to the top NAND gate, is 0 which will keep Q = 1 as represented at time t_1. At time t_2, we reset the latch by making R' = 0. Now, Q' goes to 1 and this will force Q to go to a_0. If we de-assert R' so that we have S' = R' = 1, this time the latch will remain at the reset state as shown at time t3.

Notice the 2 times (at t_1 and t_3), when both S' and R' are de-asserted. At t_1, Q is at a_1, whereas, at t_3, Q is at a0. When both the inputs are de-asserted, the SR latch maintains its previous state. Previous to t_1, Q has the value 1, so at t1, Q remains at a1. Likewise previous to t_3, Q has the value 0, so at t_3, Q remains at a_0.

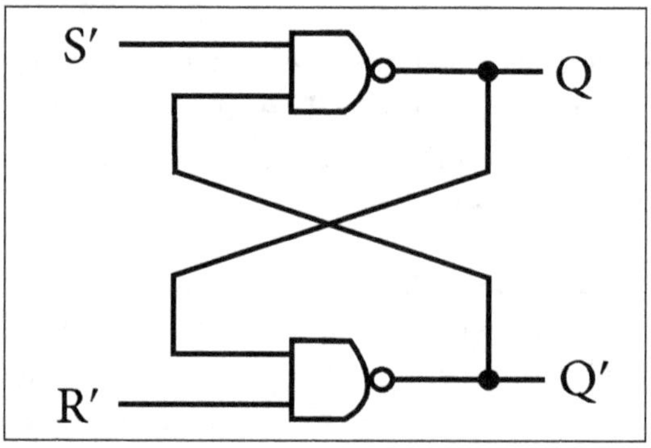

SR latch circuit using NAND gates.

Truth table:

S	R	Q	Q_{next}	$Q_{next'}$
0	0	X	1	1
0	1	X	1	0
1	0	X	0	1
1	1	0	0	1
1	1	1	1	0

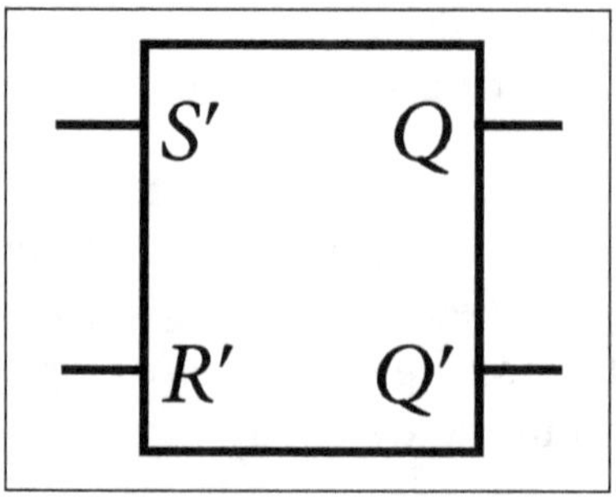

Logic symbol.

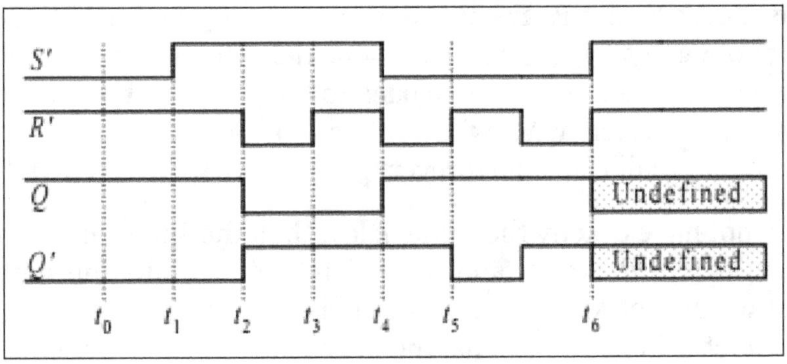

Timing diagram.

If both S' and R' are asserted, then both Q and Q' are equal to 1 as shown at time t_4. If one of the input signals is de-asserted earlier than the other, the latch will end up in the state forced by the signal that was de-asserted later as represented at time t_5. At t_5, R' is de-asserted first, so the latch goes into the normal set state with Q = 1 and Q' = 0.

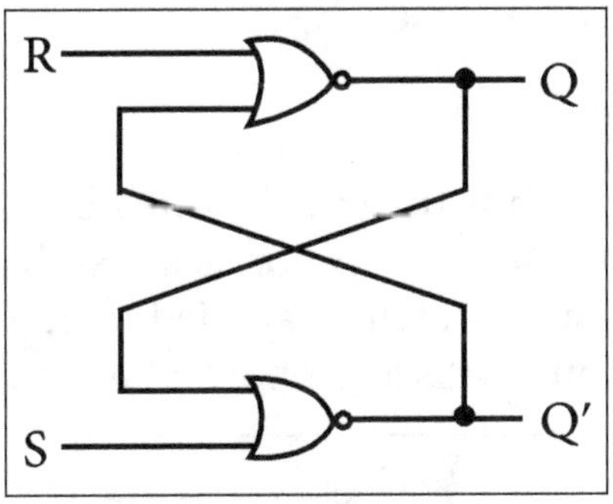

Circuit using NOR gates.

Truth table:

S	R	Q	Q_{next}	$Q_{next'}$
0	0	0	0	1
0	0	1	1	0
0	1	x	0	1
1	0	x	1	0
1	1	x	0	0

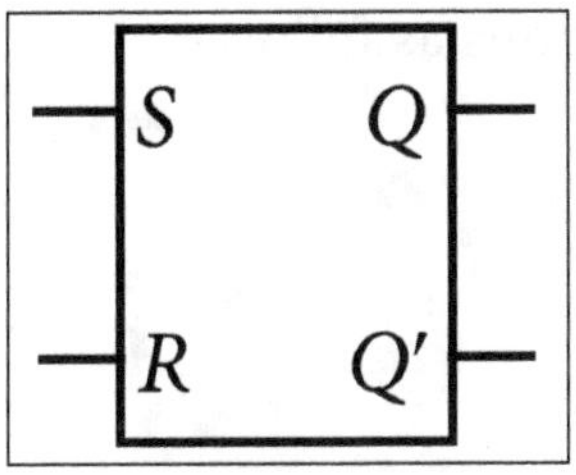

Logic symbol.

The issue exists if both S' and R' are de-asserted at exactly the same time as represented at time t6. If both gates have exactly the same delay then they will both output a_0 at exactly the same time. Feeding the zeros back to the gate input will create a1, again at exactly the same time, which again will create a_0 and so on. This oscillating behavior, called the critical race, will continue forever.

If the 2 gates do not have exactly the same delay, then the situation is similar to de-asserting one input before the other and so the latch will go into one state or the other. However, since we do not know which is the faster gate, therefore, we do not know which state the latch will go into. Thus, the latch's next state is undefined. In order to avoid this in deterministic behavior, we should make sure that the 2 inputs are never de-asserted at the same time.

Note that both of them can be de-asserted, but just not at the same time. In practice, this is guaranteed by not having both of them asserted. Another reason is, why we do not want both inputs to be asserted is that when they are both asserted, Q is equal to Q', but we usually want Q to be the inverse of Q'.

Flip-Flops

In electronics, a flip-flop or latch is the circuit which has two stable states and can be used to store the state information. A flip-flop is a bistable multi vibrator. The circuit can be made to change the state by signals applied to one or more control inputs and that will have one or two outputs. It is an important element of such circuits.

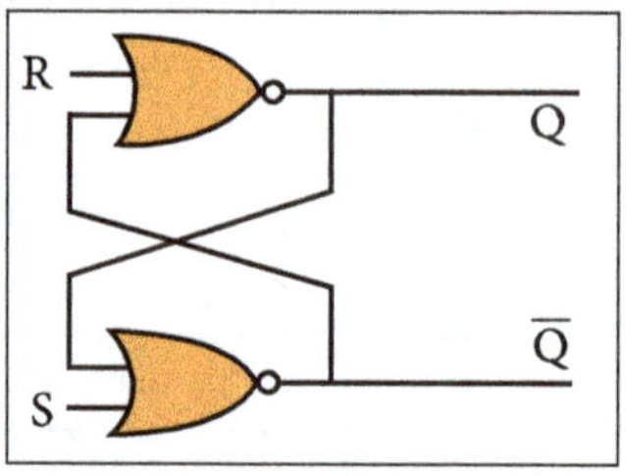

Latch or Flip-flop.

The basic unit for storage is flip flop. It maintains its output state either at 1 or 0 till it is directed by an input signal to change its state.

Different Types of Flip-Flop

There are different types of flip-flops as follows:

- SR flip-flop.
- D flip-flop.
- JK flip-flop.
- T flip-flop.

Sr Flip-Flop

The circuit remains in a particular output state indefinitely until something is done to change its output status. If the LOW and HIGH outputs are regarded as 0 and 1 respectively, then the output can either be a 0 or a 1.

Since either a 0 or a 1 can be held indefinitely until the circuit is appropriately triggered to go to the other state, the circuit is said to have memory. It is capable of storing one binary digit or one bit of digital information.

Also, in the functioning of the bistable multivibrator circuit, when one of the transistors is in saturation, the other will be in cut-off.

This implies that if we had the outputs from the collectors of both the transistors, then the two outputs would be complementary.

In the flip-flops of various types that are available in IC form, provides complementary outputs usually designated as Q and Q'.

The R-S flip-flop is the basic flip-flop. The letters S and R stands for SET and RESET.

When the flip-flop is SET, its Q output goes to state 1 and when it is RESET it goes to state 0. The Q output is the complement of the Q output at all times.

Operation of RS Flip-Flop

- When the R input is low and the S input is high, the Q output of flip-flop is set.
- When R input is high and S input is low, the Q output of the flip-flop is reset.
- When both the inputs R and S are high, the output is unpredictable.
- When both the inputs R and S are low, the output does not change.

Truth Table:

			NAND Gate		
S	R	Q(t + 1)	A	B	Y
0	0	Indeterminate	0	0	1
0	1	Set	0	1	1
1	0	Reset	1	0	1
1	1	NC	1	1	0

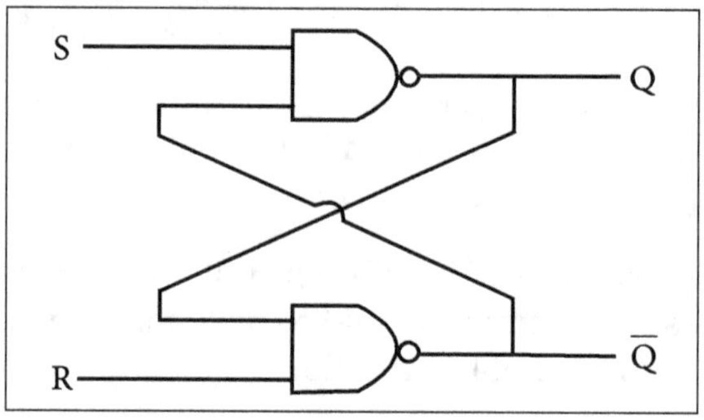

Logic diagram of SR flip-flop.

Jk Flip-Flop

The J-K flip-flop behaves in the same fashion as an R-S flip-flop except for one entry in the function table.

In case of R-S flip-flop, its input combination S = R = 1 and the input combination S = R = 0 are prohibited.

In the case of the J-K flip-flop with active HIGH inputs, the output of the flip-flop toggles, (i.e.), it goes to the other state, for J = K = 1.

The output toggles for J = K = 0 in the case of the flip-flop having a active LOW input. Thus, the J-K flip-flop can overcome the problem of a forbidden input combination of the R-S flip-flop.

The corresponding Karnaugh maps are shown in below figure for the characteristics table.

The characteristic equations is given by,

$$Q_{N+1} = J.\overline{Q}_N + \overline{K}.Q_N$$

$$Q_{N+1} = \overline{J}.\overline{Q}_N + K.Q_N$$

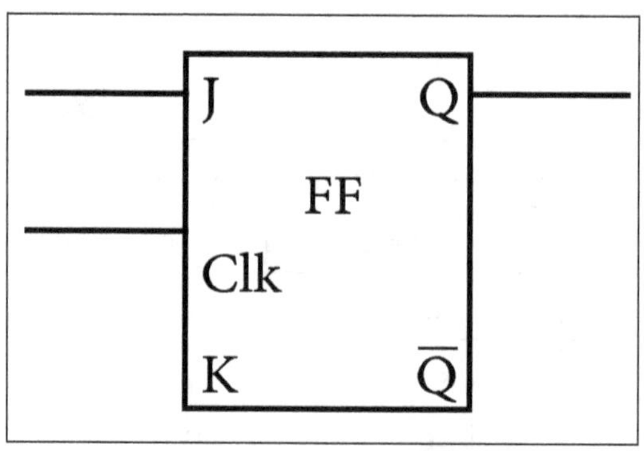

(a) JK flip flop with active high inputs.

Truth table:

Operation mode	J	K	Clk	Q_{n+1}
SET	1	0	1	1
RESET	0	1	1	0
NO CHANGE	0	0	1	Qn
TOGGLE	1	1	1	$\bar{Q}_N$

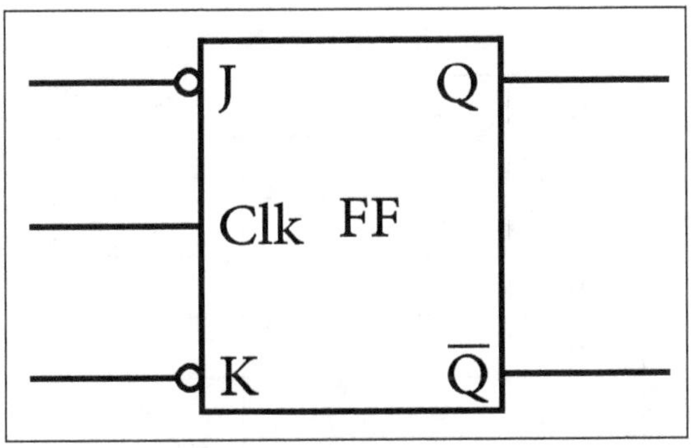

(b) JK flip flop with active low inputs.

Truth table:

Operation mode	J	K	Clk	Q_{n+1}
SET	0	1	1	1
RESET	1	0	1	0
NO CHANGE	1	1	1	Q_n
TOGGLE	0	0	1	$\bar{Q}_N$

Jk Flip-Flop is a Universal Flip-Flop

When configured in various ways, JK flip flop is capable of operating like most other types of flip - flops.

Operation of Jk Flip-Flop

- When the K input is low and the J input is high, the Q output of flip-flop is set.
- When the K input is high and the J input is low, the Q output of flip-flop is reset.
- When both the inputs K and J are low, the output does not change.
- When both the inputs K and J are high, it is possible to set or reset the flip-flop (i.e.) the output toggle on the next positive clock edge.

State Diagram

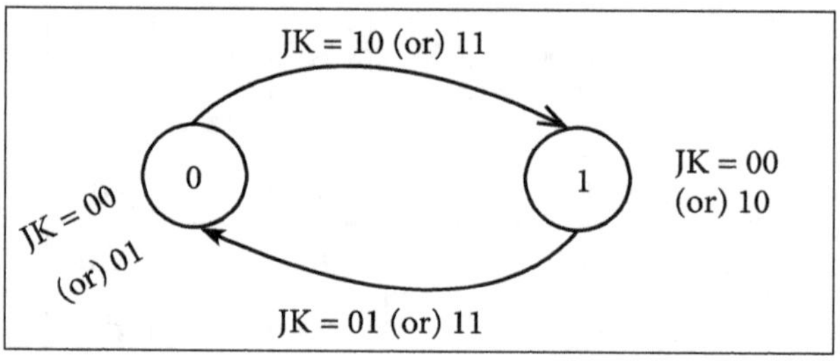

State table:

Q	J	K	Q(t + 1)
0	0	0	0
0	0	1	0
0	1	0	1
0	1	1	1
1	0	0	1
1	0	1	0
1	1	0	1
1	1	1	0

Counter Using Jk Flip Flop

Q_2	Q_1	Q_0
0	0	0
0	0	1
0	1	1
1	1	1
1	1	0
1	0	0
0	0	0

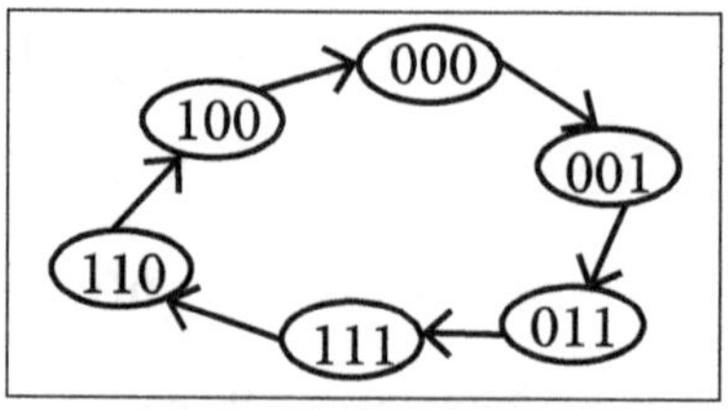

State Diagram.

State table:

PS			NS			JK Input					
Q_2	Q_1	Q_0	Q_{0+1}	Q_{2+1}	Q_{1+1}	J_2	K_2	J_1	K_1	J_0	K_0
0	0	0	1	0	0	0	X	0	X	1	X
0	0	1	1	0	1	0	X	1	X	X	0
0	1	0	0	0	0	0	X	X	1	0	X
0	1	1	1	1	1	1	X	X	0	X	0
1	0	0	0	0	0	X	1	0	X	0	X
1	0	1	0	0	0	X	1	0	X	X	1
1	1	0	0	1	0	X	0	X	1	0	X
1	1	1	0	1	1	X	0	X	0	X	1

K–Map:

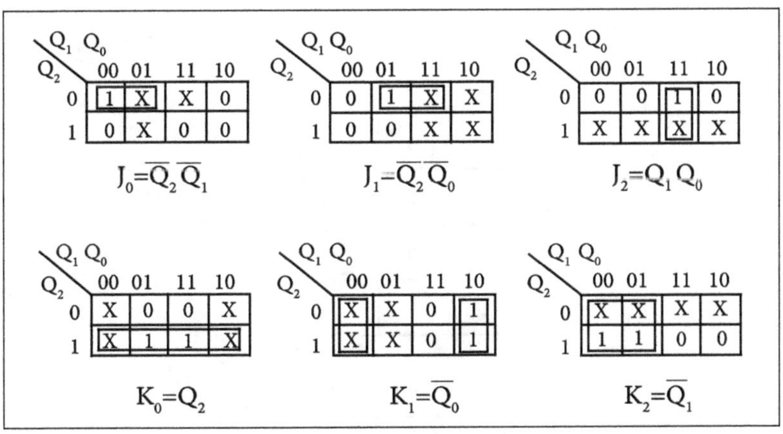

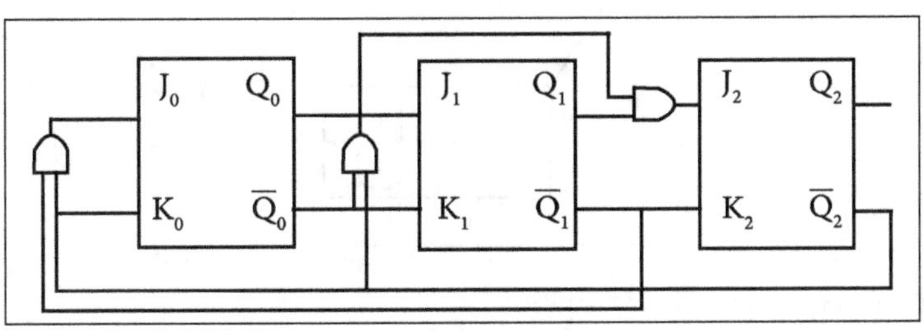

Logic Diagram.

D Flip-Flop

A D flip-flop also known as a delay flip-flop can be used to provide temporary storage of one bit of information. Figure below shows the symbol of the circuit and the function table of a negative edge-triggered D flip-flop. When the clock is active, the data bit (0 or 1) present at the D input is transferred to the output. In the D flip-flop, the data transfer from D input to Q output occurs on the negative-going transition of the clock input. D input can acquire new status.

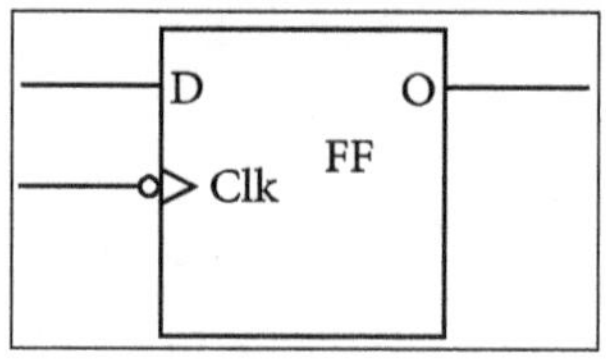

Symbol

Function table:

D	Clk	Q
0	↓	0
1	↓	1

Truth table:

Q_n	D	Q_{n+1}
0	0	0
0	1	1
1	0	0
1	1	1

K-Map:

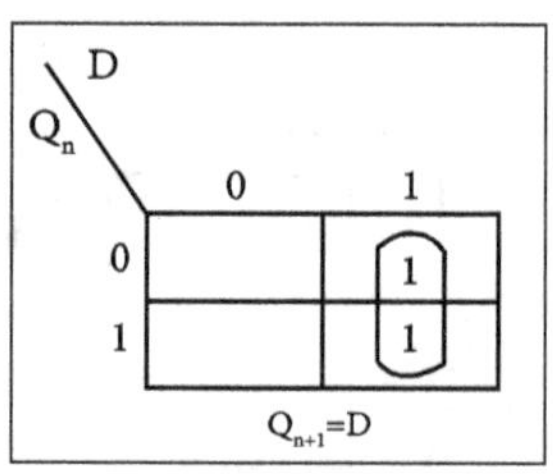

Operation of D Flip-Flop

In D flip-flop during the occurrence of clock pulse if D=1, the output Q is set and if D=0, the output is reset.

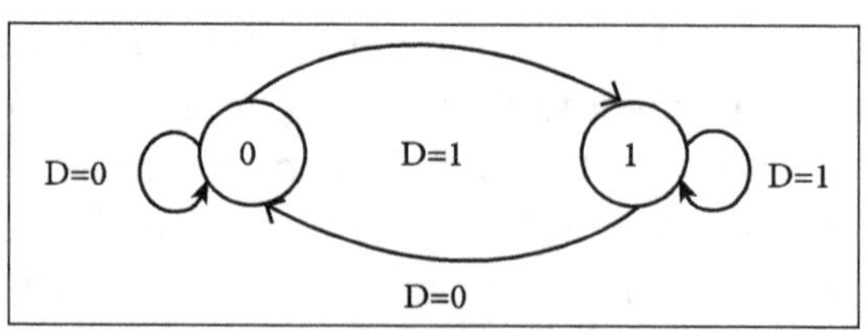

State diagram.

State table:

Q	D	Q(t+1)
0	0	0
0	1	1
1	0	0
1	1	1

Flip-Flop Excitation Tables for D Flip-Flop

In D flip-flop, its next state is always equal to the D input and it is independent of the present state. D is 0 if Q_{n+1} has to 0 and if Q_{n+1} has to be 1 regardless the value of Q_n.

T Flipflop

The output of a toggle flip-flop also called a T flip-flop changes state every time it is triggered at its T input known as the toggle input. That is, its output becomes 1, if it was 0 and 0 if it was 1.

The characteristic equations can be written as follows:

$$Q_{N+1} = T.\overline{Q}_N + \overline{T}.Q_N$$
$$Q_{N+1} = \overline{T}.\overline{Q}_N + T.\overline{Q}_N$$

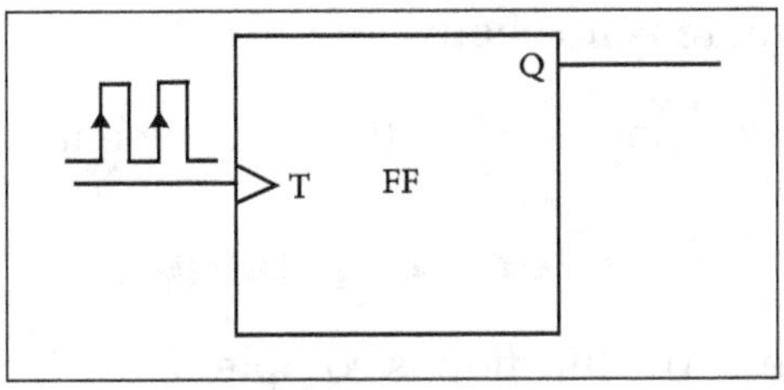

T flip-flop.

T	Q_n	Q_{n+1}
↑	0	1
↑	1	0

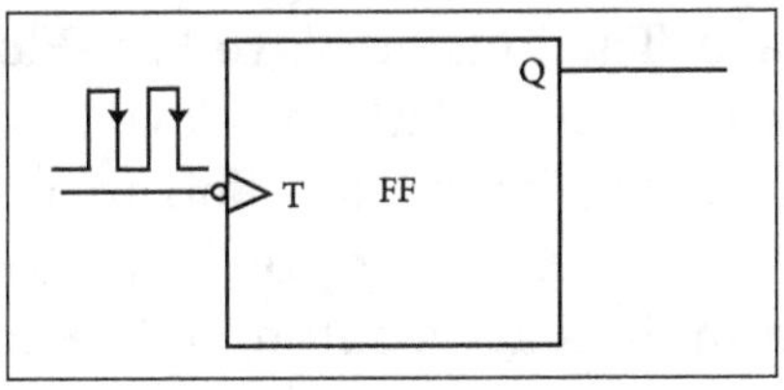

T flip-flop.

T	Q_n	Q_{n+1}
↓	0	1
↓	1	0

Truth table:

Q_n	T	Q_{n+1}
0	0	0
0	1	1
1	0	1
1	1	0

Q_n	T	Q_{n+1}
0	0	1
0	1	0
1	0	0
1	1	1

Operation of T Flip-Flop

T flip-flop is also known as Toggle flip-flop.

When the value of T = 0, there is no change in the output.

When the value of T = 1, the output switch to the complement state (i.e) the output toggles.

Flip-Flop Excitation Tables for T Flip-Flop

When input T = 1 the state of the flip-flop is complemented.

When T = 0, the state of the Flip-flop remains unchanged. Therefore, for 00 and 11 transitions T must be 0 and for 01 and 10 transitions T must be 1.

Master-Slave

A master-slave flip-flop consists of two flip-flops where one circuit serves as a master and the other serves as a slave. The Master-Slave Flip-Flop is basically two gated SR flip-flops connected together in a series with the slave having an inverted clock pulse. The outputs from Q and $\bar{Q}$ from the "Slave" flip-flop are given back to the inputs of the "Master" with the outputs of the "Master" flip flop being connected to the two inputs of the "Slave". This feedback from the slave's output to the master's input gives the characteristic toggle of the JK flip flop as shown in the below figure.

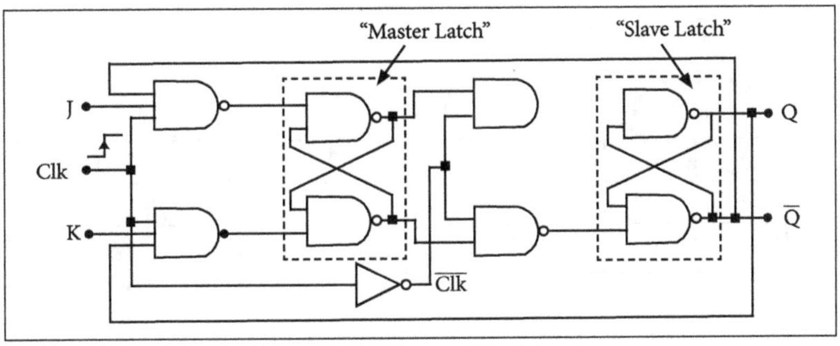

Master-Slave JK Flip Flop.

The input signals J and K are connected to the gated "master" SR flip flop "locks" the input condition while the clock input is "HIGH" at logic level "1". As the clock input of the "slave" flip flop is the complement of the "master" clock input, the "slave" SR flip flop doesn't toggle. The outputs from the "master" flip flop are only seen by the gated "slave" flip flop when the clock input goes "LOW" to the logic level "0".

When the clock is at "LOW" level, the outputs from the "master" flip flops are latched and any further changes to its inputs are not considered. The gated "slave" flip flop now responds to the state of its inputs passed over by the "master" section. Then, on the "Low to High" transition, the inputs of the "master" flip flop are fed to the gated inputs of the "slave" flip flop and on the "High to Low" transition, the same inputs are reflected on the output of the "slave" making this type of flip flop pulse or edge-triggered. Then, the circuit accepts input data when the clock signal is "HIGH" and passes the data to the output on the falling-edge of the clock signal.

In other words, the Master-Slave JK Flip flop is a "Synchronous" device as it passes only the data with the timing of the clock signal.

Application of Master Slave FF

- Frequency dividers.
- Shift registers.
- Parallel data storage.
- Counters.

Problems

1. Let us consider that the following sequence is to be realized by a counter consisting of 3 JK Flip flops.

A_1	0	0	0	0	1	1	0
A_2	0	1	1	0	0	1	0
A_3	0	1	0	1	1	0	0

Solution:

Given:

A_1	0	0	0	0	1	1	0
A_2	0	1	1	0	0	1	0
A_3	0	1	0	1	1	0	0

Truth table:

Present State			**Next State**			**Flip-Flop Inputs**					
A_1	A_2	A_3	A_1	A_2	A_3	JA_1	KA_1	JA_2	KA_2	JA_3	KA_3
0	0	0	0	1	1	0	X	1	X	1	X
0	1	1	0	1	0	0	X	X	0	X	1
0	1	0	0	0	1	0	X	X	1	1	X
0	0	1	1	0	1	1	X	0	X	X	0
1	0	1	1	1	0	X	0	1	X	X	1
1	1	0	0	0	0	X	1	X	1	0	X

K – map:

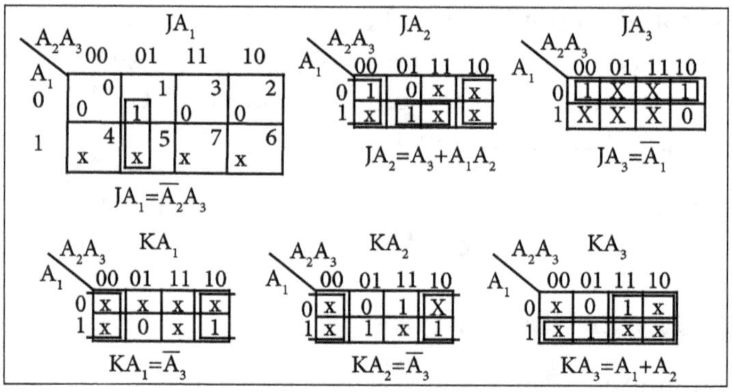

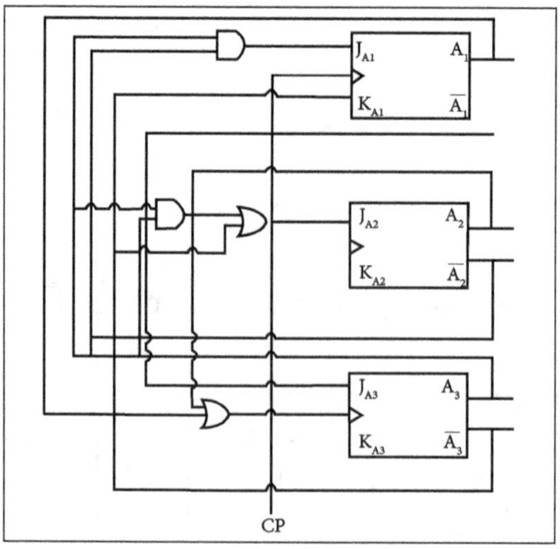

2.3.1 Conversion of Flip-Flops

JK Flip Flop to SR Flip-Flop

S and R will be the external inputs to J and K. As shown in the logic diagram below, J and K values will be the outputs of the combinational circuit. Thus, the values of J and K have to be obtained in terms of S, R and Qp. The logic diagram is shown below.

A conversion table is to be written using S, R, Qp, Qp+1, J and K. For two inputs S and R, eight combinations can be made. For each combination, its corresponding Qp+1 outputs are found out. The outputs for the combinations of S=1 and R=1 are not permitted for an SR flip flop. Hence, the outputs are considered as invalid and the J and K values are taken as "don't cares".

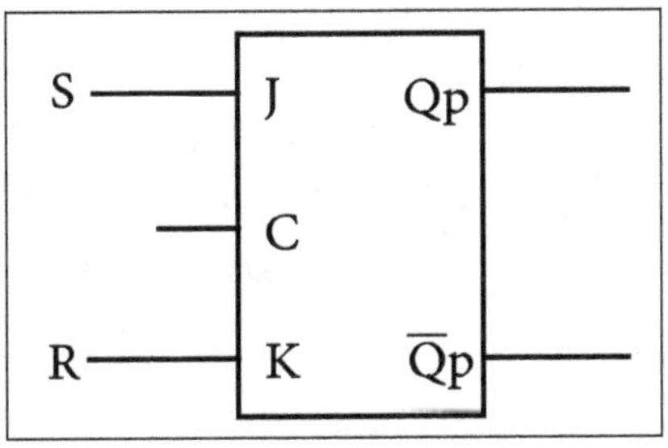

Logic diagram.

Conversion table:

S-R inputs		Outputs		J-K inputs	
S	R	Q_P	Q_{P+1}	J	K
0	0	0	0	0	X
0	0	1	1	X	0
0	1	0	0	0	X
0	1	1	0	X	1
1	0	0	1	1	X
1	0	1	1	X	0
1	1	Invalid		Don't care	
1	1	Invalid		Don't care	

K-Map:

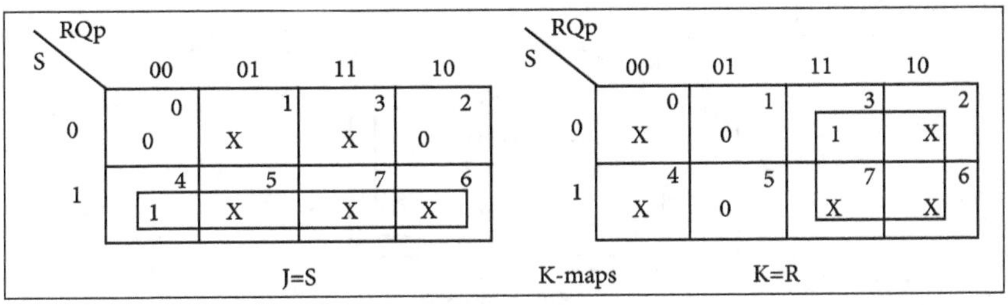

SR Flip Flop to D Flip-Flop

As shown in the figure, S and R are the actual inputs of the flip flop and D is the external input of the flip flop. The four combinations, the logic diagram, conversion table and the K-map for S and R in terms of D and Q_p are shown below.

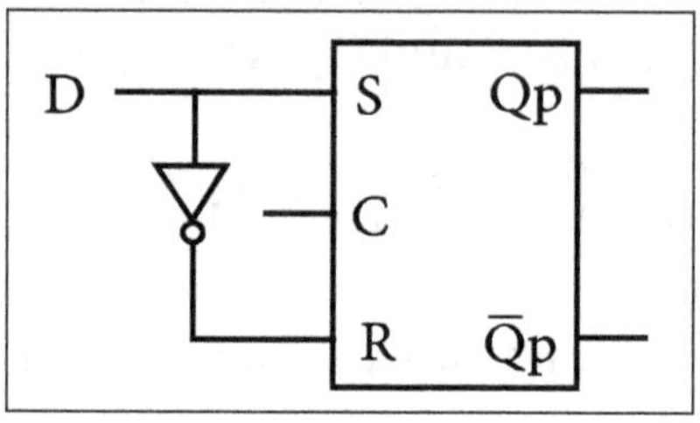

Logic diagram.

Conversion table:

D input	Outputs		S-R inputs	
	Q_p	Q_{p+1}	S	R
0	0	0	0	x
0	1	0	0	1
1	0	1	1	0
1	1	1	x	0

K-Map:

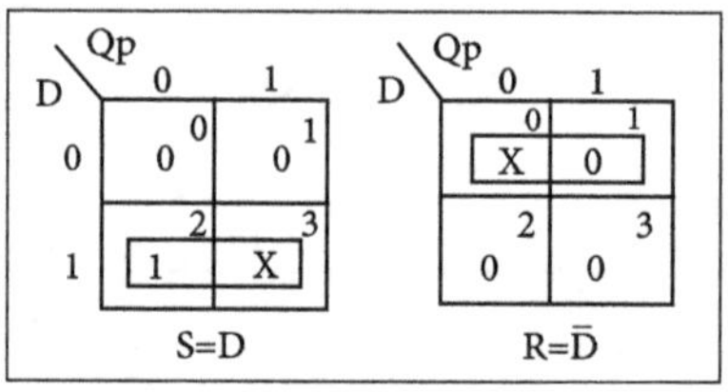

D Flip Flop to SR Flip-Flop

D is the actual input of the flip flop and S and R are the external inputs. Eight possible combinations are achieved from the external inputs S, R and Qp. Since the combination of S=1 and R=1 are invalid, values of Q_{p+1} and D are considered as "don't cares". Logic diagram showing the conversion from D to SR and the K-map for D in terms of S, R and Qp are shown below.

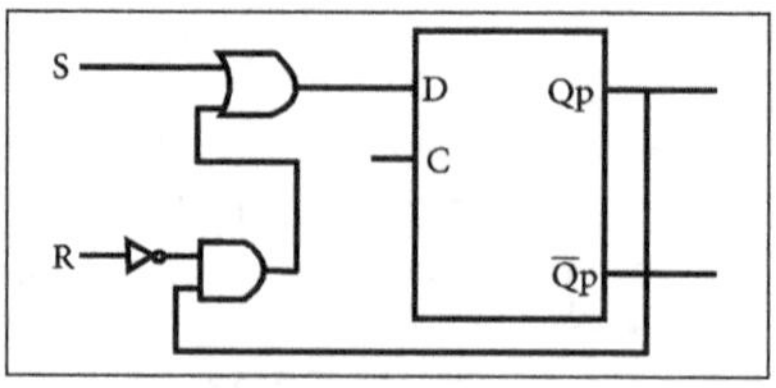

Logic diagram.

Conversion table:

S-R inputs		Outputs		D inputs
S	R	Q_P	Q_{P+1}	
0	0	0	0	0
0	0	1	1	1
0	1	0	0	0
0	1	1	0	0
1	0	0	1	1
1	0	1	1	1
1	1	Invalid		Don't care
1	1	Invalid		Don't care

K-Map:

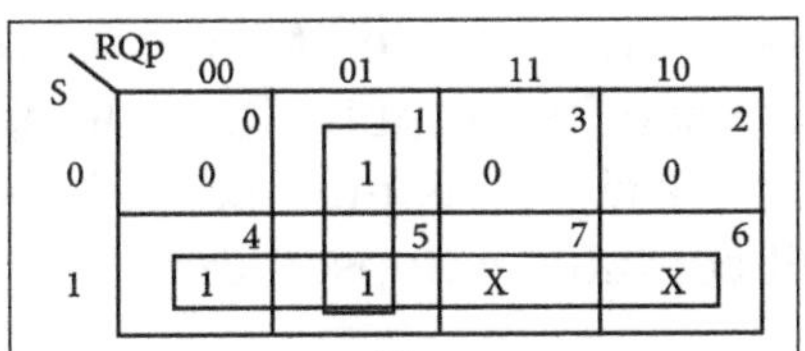

JK Flip Flop to T Flip-Flop

J and K are the actual inputs of the flip flop and T is taken as the external input for conversion. Four combinations are produced with T and Qp. J and K are expressed in terms of T and Qp.

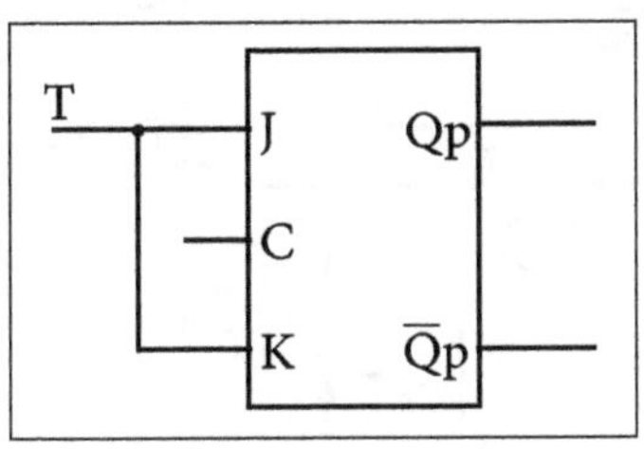

Logic diagram.

Conversion table:

T input	Outputs		J-K inputs	
	Qp	Q_{p+1}	J	K
0	0	0	0	x
0	1	1	x	0
1	0	1	1	x
1	1	0	x	1

K-Map:

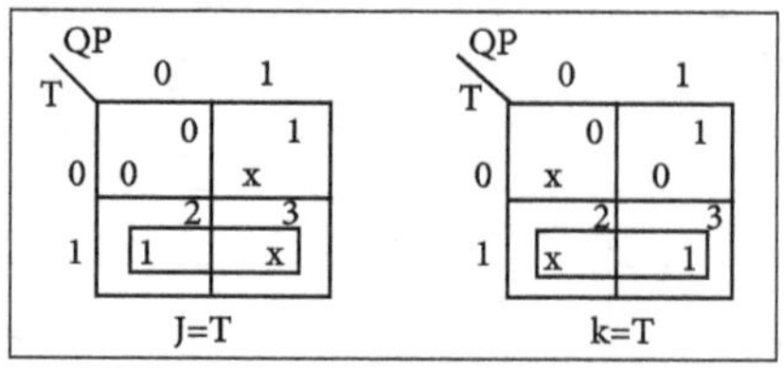

2.4 Analysis of Clocked Sequential Circuits and Mealy and Moore Models of Finite State Machines

Analysis of Synchronous Sequential Circuits

The analysis of a synchronous sequential circuit is the process of determining the functional relation that exists between its outputs, its inputs and its internal states. The contents of all the flip-flops in the circuit combined will determine the internal state of the circuit. Thus, if the circuit contains n flops, it can be in one of the r states.

Knowing the present state of the circuit and the input values at any time t, we should be able to derive its next state (i.e., the state at time t + 1) and the output produced by the circuit at t. A sequential circuit can be described completely by a state table that is very similar to the ones shown for flip-flops. For a circuit with n flip-flops, there will be 2n rows in the state table.

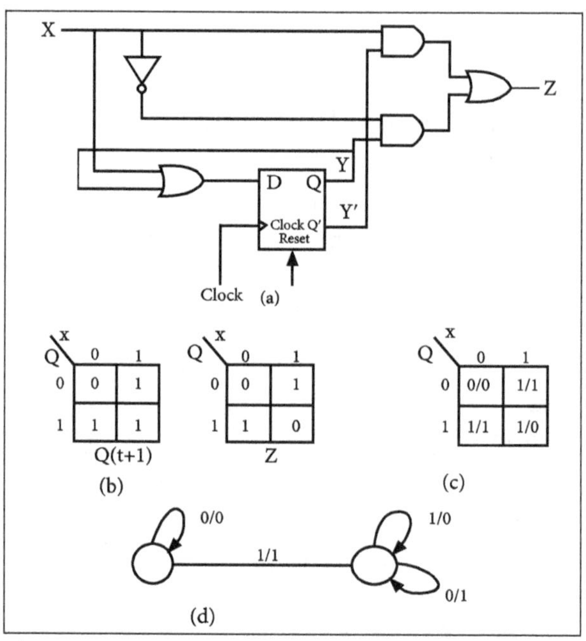

Sequential circuit analysis: (a) Circuit; (b) next-state and output tables; (c) transition table; (d) state diagram.

If there are m inputs to the circuit, there will be 2_m columns in the state table. At the intersection of each row and column, the next-state and the output information are recorded. A state diagram is a graphical representation of the state table, in which each state is represented by a circle and the state transitions are represented by arrows between the circles. The input combination that brings about the transition and the corresponding output information are shown on the figure.

Analyzing a sequential circuit thus corresponds to generate the state table or the state diagram for the circuit. The state table or state diagram can be used to determine the output sequence generated by the circuit for a ten input sequence if the initial state is known.

It is important to note that for proper operation, a sequential circuit must be in its initial state before the inputs to it can he applied. Usually the power-up circuits are used to initialize the circuit to the appropriate state when the power is turned on.

Mealy and Moore Models

A synchronous sequential circuit is a system whose behavior can be defined from the knowledge of its signals at discrete instants of time.

Secondary Variables

The delay elements are used to provide a short term memory for the sequential circuit. The present state and next state variables in asynchronous sequential circuits are called secondary variables.

Present State and Next State

The information stored in the memory elements at any given time defines the present state of the sequential circuit. The present state and the external inputs determine the outputs and the next state of the sequential circuits.

General Model

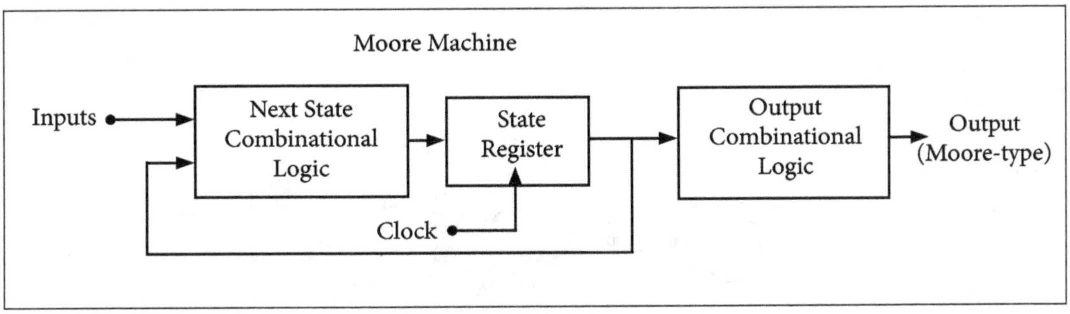

Moore machine.

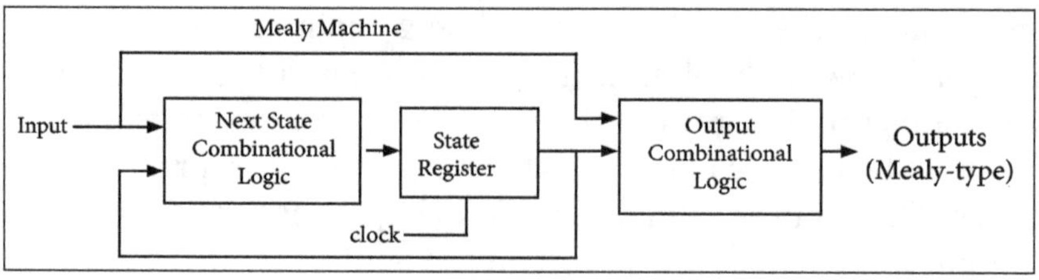

Mealy design.

State Transition Diagram of Moore and Mealy Model

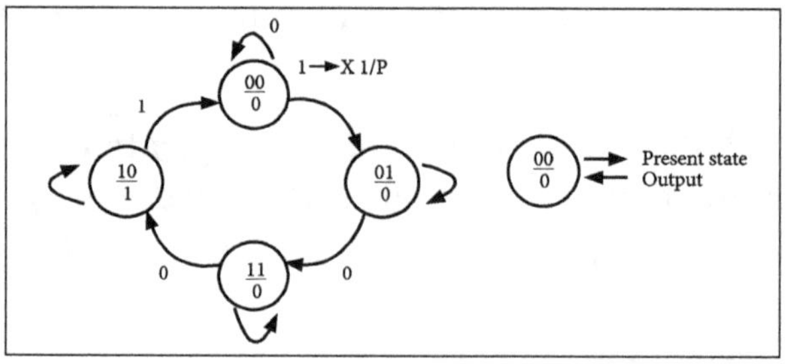

State diagram of a Moore circuit.

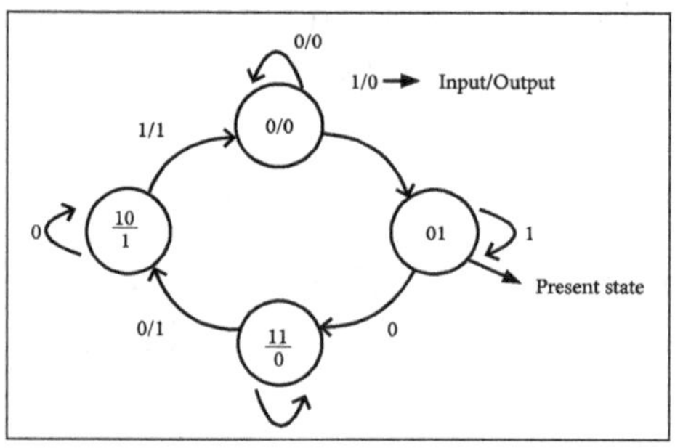

State diagram of Mealy circuit.

Problems

1. Let us draw the state transition diagram of a sequence detector circuit that detects '1010' from input data stream using Moore model and Mealy model.

Solution:

Given:

1010

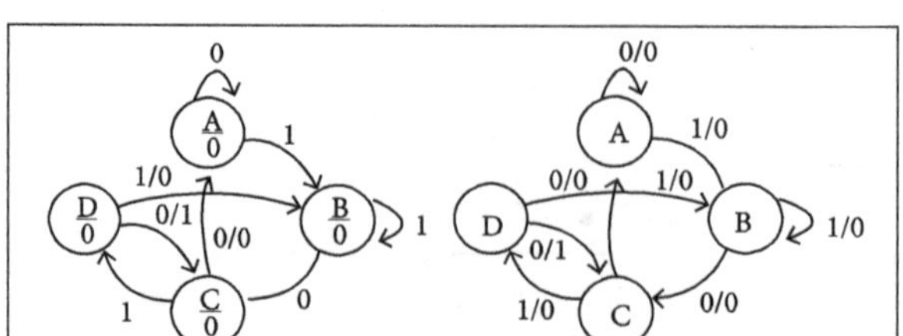

Moore model Mealy model.

State Table:

PS	NS		Y	
	X = 0	X = 1	X = 0	X = 1
A	A	B	0	0
B	C	B	0	0
C	A	D	0	0
D	C	B	1	0

2.5 Binary Counters

Binary Counter

A register that goes through a prescribed sequence of states upon the application of input pulses is called counter. A counter that follows the binary number sequence is called a binary counter.

Lockout Condition

- A counter is supposed to follow the sequence of the derived states. If it enters into an unused or unwanted state, then it is expected to return back to a desired state.
- If the next state of an unwanted state is again an unwanted state, then the counter is said to be in the lockout condition.

4-Bit Decode Counter

Decode counter has a count sequence from 0000(0) to 1001(9). After 1001 state, it must recycle to 0000 state.

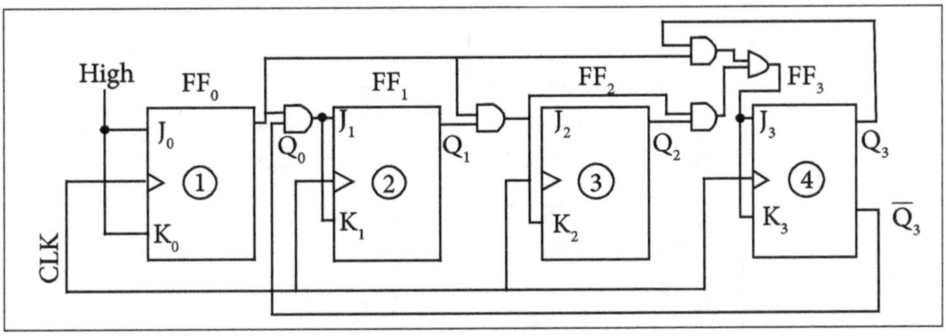

4 - bit Synchronous Decode Counter.

This counter requires four flip - flops and AND / OR logic gates.

$J_0 = K_0 = 1$

$J_1 = K_1 = Q_0Q_3$

$J_2 = K_2 = Q_0Q_1$

$J_3 = K_3 = Q_0Q_1Q_2 + Q_0Q_3$

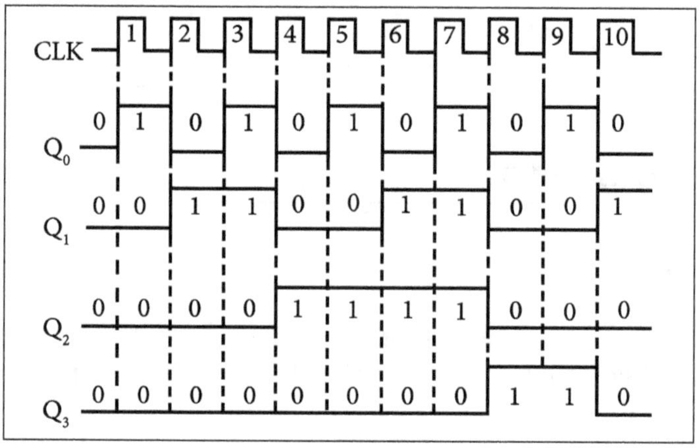

Timing Diagram.

Design of 3-Bit Binary Up-Down Counter

An UP/DOWN counter is one that is capable of progressing in increasing order or decreasing order through a certain sequence. The state table is given as follows.

CP	UP	Q_c	Q_b	Q_a	DOWN
0		0	0	0	
1		0	0	1	
2		0	1	0	
3		0	1	1	
4		1	0	0	
5		1	0	1	
6		1	1	0	
7		1	1	1	

The design of 3-bit UP/DOWN counter is made by using T-flipflop.

Excitation Table:

Input UP/DOWN UD	Present State			Next State			Flip-Flop Inputs		
	Q_c	Q_b	q_a	Q_{c+1}	Q_{b+1}	q_{a+1}	T_c	T_b	T_a
0	0	0	0	1	1	1	1	1	1
0	0	0	1	0	0	0	0	0	1
0	0	1	0	0	0	1	0	1	1
0	0	1	1	0	1	0	0	0	1
0	1	0	0	0	1	1	1	1	1
0	1	0	1	1	0	0	0	0	1
0	1	1	0	1	0	1	0	1	1
0	1	1	1	1	1	0	0	0	1
1	0	0	0	0	0	1	0	0	1
1	0	0	1	0	1	0	0	1	1
1	0	1	0	0	1	1	0	0	1
1	0	1	1	1	0	0	1	1	1
1	1	0	0	1	0	1	0	0	1
1	1	0	1	1	1	0	0	1	1
1	1	1	0	1	1	1	0	0	1
1	1	1	1	0	0	0	1	1	1

K-Map Simplification:

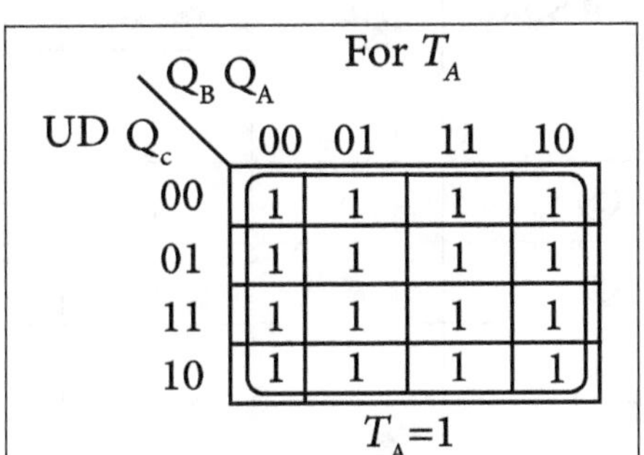

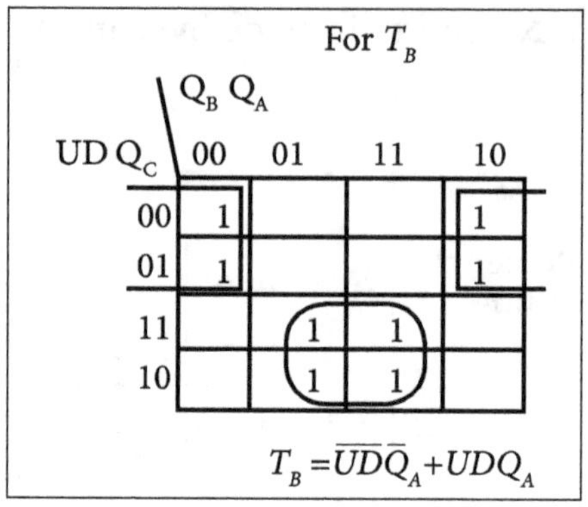

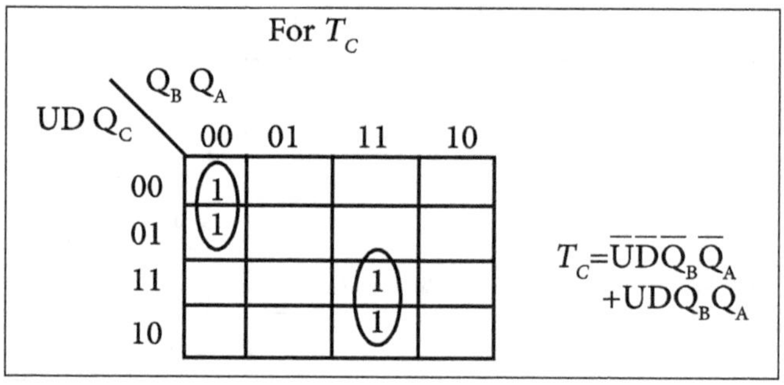

State Diagram and State Table

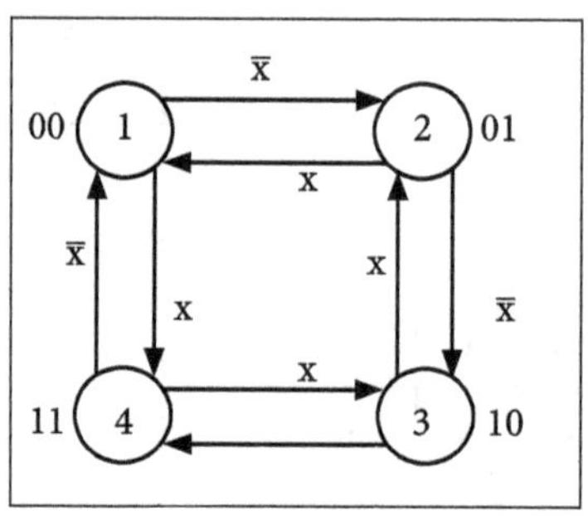

State diagram.

State table:

Present State	Next	State	
①	$\overline{x}$	x	$Z_1 Z_2$
②	2	4	00
③	3	1	01
④	4	2	10
	1	3	11

Logic Diagram

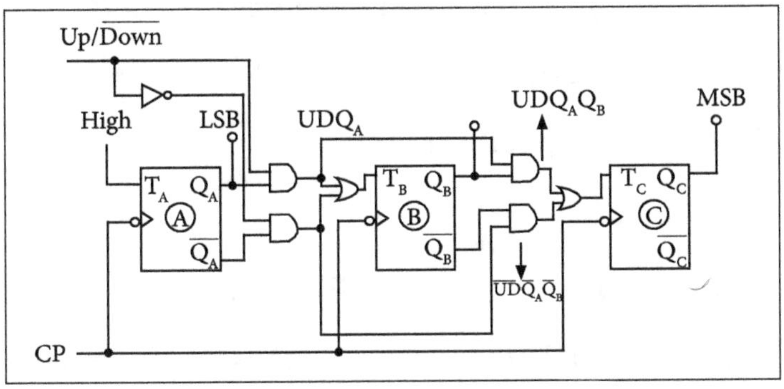

2.5.1 Principle and Design of Synchronous and Asynchronous Counters

A synchronous counter is one whose output bits change state simultaneously, with no ripple. The only way we can build the synchronous counter circuit from J-K flip-flops is to connect all the clock inputs together, so that each and every flip-flop receives the exact same clock pulse at the exact same time.

In the synchronous counters, the clock inputs of all the flip-flops are connected together and are triggered by the input pulses. Thus, all the flip-flops change state simultaneously. The number of states in a 3-bit ring counter is 8.

The states are as follows:

C_P	Q_C	Q_B	Q_A
0	0	0	0
1	0	0	1
2	0	1	0
3	0	1	1
4	1	0	0
5	1	0	1
6	1	1	0
7	1	1	1

The procedure to design a synchronous counter is listed here:

- Obtain the truth table of the logic sequence for intended counter to be designed. Alternatively obtain the state diagram of the counter.
- Determine number and type of flip-flop to be used.
- From the excitation table, determine its next state logic.

- From the output state, use Karnaugh map for simplification to derive the circuit output functions and flip-flop output functions.
- Draw the logic circuit diagram.
- Simulate the circuit using software.
- Build the circuit.

Synchronous BCD counter using JK Flip-Flop

To design a BCD or Decade (MOD-10) counter that has ten states, i.e., 0 to 9. The number of flip flop required is four. Let us assume that the MOD-10 counter has ten states.

Step 1: State Diagram

Now the State diagram for the MOD-10 counter can be drawn as shown. Here it is assumed that the state transition from one state to another takes place when the clock pulse is asserted. When the clock is un-asserted, the counter remain in the present state.

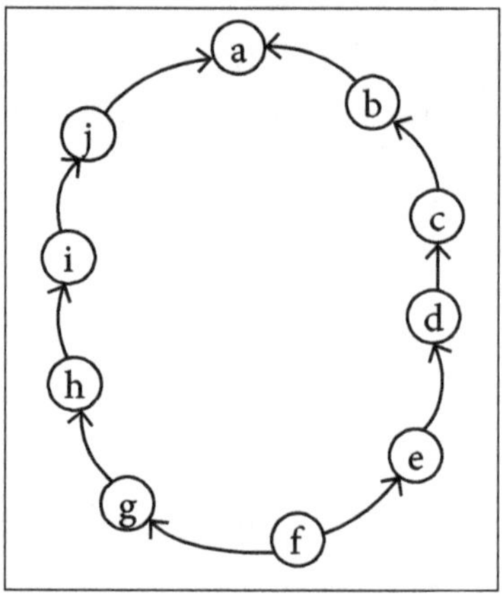

State diagram of BCD counter.

Step 2: State Table

From the above state diagram, one can draw the PS-NS table as:

PS	NS
a	b
b	c
c	d
d	e
e	f
f	g

g	h
h	i
i	j

The above state table does not have any redundant state because no two states are equivalent. So there is no modification required in the above state table.

Step 3: State Assignment

Let us assign four state variables to these states a, b, c, d, e, f, g, h, i and j as follows: a = 0000, b = 0001, c = 0010, d = 0011, e = 0100, f = 0101, g = 0110, h = 0111, i = 1000 and j = 1001. Then the above PS- NS table can be modified as shown:

PS				**NS**			
q_3	q_2	q_1	q_0	Q_3	Q_2	Q_1	Q_0
0	0	0	0	0	0	0	1
0	0	0	1	0	0	1	0
0	0	1	0	0	0	1	1
0	0	1	1	0	1	0	0
0	1	0	0	0	1	0	1
0	1	0	0	0	1	0	1
0	1	0	1	0	1	1	0
0	1	1	1	1	0	0	0
1	0	0	0	1	0	0	1
1	0	0	1	0	0	0	0

Step 4: Excitation Table

The excitation tables having entries for flip flop inputs (J_3K_3, J_2K_2, J_1K_1 and J_0K_0) can be drawn.

PS				**NS**				**Excitation Inputs**							
q_3	q_2	q_1	q_0	Q_3	Q_2	Q_1	Q_0	J_3	K_3	J_2	K_2	J_1	K_1	J_0	K_0
0	0	0	0	0	0	0	1	0	d	0	d	0	d	1	d
0	0	0	1	0	0	0	0	0	d	0	d	1	d	d	1
0	0	1	0	0	0	1	1	0	d	0	d	d	0	1	d
0	0	1	1	0	0	1	0	0	d	1	d	d	1	d	1
0	1	0	0	0	1	0	1	0	d	d	0	0	d	1	d

0	1	0	1	0	1	0	0	0	d	d	0	1	d	d	1
0	1	1	0	0	1	1	1	0	d	d	0	d	0	1	d
0	1	1	1	0	1	1	0	1	d	d	1	d	1	d	1
0	0	0	0	0	0	0	1	1	0	0	d	0	d	1	d
0	0	0	1	0	0	0	0	d	1	0	d	0	d	d	1
0	0	0	0	0	0	0	1	d	0	d	0	0	0	1	d
0	0	0	0	0	0	0	0	0	1	d	0	0	0	d	1

Step 5: Excitation Maps

The excitation maps for J_3, K_3, J_2, K_2, J_1, K_1, J_o and K_o inputs of the counter can be drawn.

K-Maps:

q_0q_1 \ q_0q_1	00	01	11	10
	0	0	d	d
01	0	0	d	d
11	0	1	d	d
10	0	0	d	d

(a) For J_3

q_0q_1 \ q_0q_1	00	01	11	10
00	d	d	d	0
01	d	d	d	1
11	d	d	d	d
10	d	d	d	d

(b) For K_3

q_0q_1 \ q_0q_1	00	01	11	10
00	0	d	d	0
01	0	d	d	0
11	1	d	d	d
10	0	d	d	d

(c) For J_2

q_0q_1 \ q_0q_1	00	01	11	10
00	d	0	d	d
01	d	0	d	d
11	d	1	d	d
10	d	0	d	d

(c) For K_2

q_0q_1 \ q_0q_1	00	01	11	10
00	0	0	d	0
01	1	1	d	0
11	d	d	d	d
10	d	d	d	d

(e) For J_1

q_0q_1 \ q_0q_1	00	01	11	10
00	d	d	d	d
01	d	d	d	d
11	1	1	d	d
10	0	0	d	d

(e) For K_2

Step 6: Schematic Diagram

Using the above excitation, the circuit diagram can be drawn.

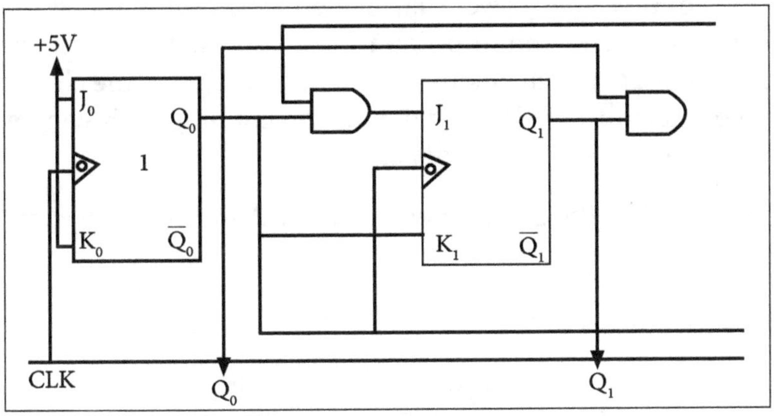

Logic diagram for BCD counter.

Comparison between Synchronous and Asynchronous Counters

Asynchronous counters	Synchronous counters
In this type of counter, flip-flops are connected in such a way that the output of first flip-flop drives the clock for the next flip-flop.	In this type, it has no connection between the output of the first flip-flop and the clock input of the next flip - flop.
All the flip-flops are not simultaneously clocked.	All the flip-flops are clocked simultaneously.

Asynchronous Ripple Counter

A ripple counter is an asynchronous counter where only the first flip-flop is been clocked by an external clock. Every subsequent flip-flops are clocked by the output of the preceding flip-flop. The asynchronous counters are also known as ripple-counters since the way in which the clock pulse ripples it through the flip-flops.

The MOD of the asynchronous counter or ripple counter is 2n if n flip-flops are used. For a 4-bit counter, the range of the count is 0000 to 1111 (2^4-1). The counter may count up or count down or count up and down depending on the input control. Thus, the count sequence usually repeats itself. During the process of counting up, the count sequence shall go from 0000, 0001, 0010, ... 1110, 1111, 0000, 0001, ... etc., whereas when counting down the count sequence shall go in the opposite manner: 1111, 1110, ... 0010, 0001, 0000, 1111, 1110, ... etc.

Thus, the complement of the count sequence counts in reverse direction. If the un-complemented output counts up, the complemented output counts down and vice versa.

There are many ways to implement the ripple counter depending on the characteristics of the flip flops used and requirements of the count sequence.

The Clock Trigger is either Positive edged or Negative edged. The flip flops used are JK or D flip-flops. The count direction takes place are up, down or up/down.

The asynchronous counters are slower than synchronous counters because of the delay in transmission of the pulses from flip-flop to flip-flop. In synchronous circuit, all the bits in the counter changes synchronously with an assertion of the clock. The examples of synchronous counters are the Ring and Johnson counter. This can be implemented using D-type flip-flops or JK-type flip-flops.

Circuit below uses 2 D flip-flops to implement a divide-by-4 ripple counter (2n = 22 = 4). It counts down.

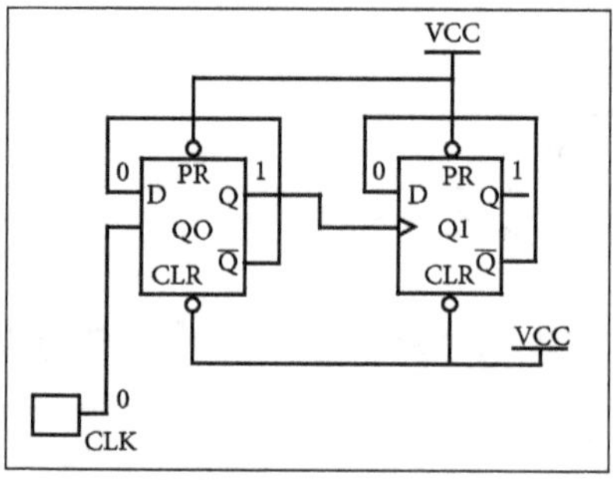

Divide-by-4 ripple counter.

State table:

Count	Q_1	Q_0
3	1	1
2	1	0
1	0	1
0	0	0

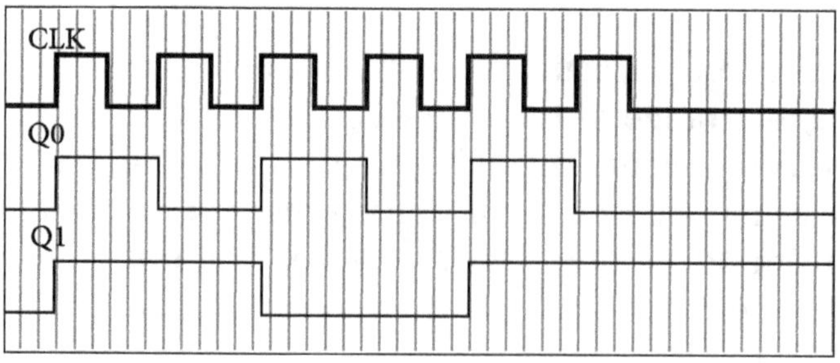

Timing diagram.

Asynchronous Up-Counter

In the up-counter, each flip-flop is triggered by the normal output of the preceding flip-flop whereas in the down-counter, each flip-flop is triggered by the complement output of the preceding flip-flop.

The operation of such counter is controlled by the up-down control input:

COUNT-UP mode				COUNT-DOWN mode			
States	Q_C	Q_B	Q_A	States	Q_C	Q_B	Q_A
0	0	0	0	7	1	1	1
1	0	0	1	6	1	1	0
2	0	1	0	5	1	0	1
3	0	1	1	4	1	0	0
4	1	0	0	3	0	1	1
5	1	0	1	2	0	1	0
6	1	1	0	1	0	0	1
7	1	1	1	0	0	0	0

For up down counting operation preceding flip-flop, sometime it requires input from output Q of first flip-flop to clock of next flip-flop for up-counting and sometimes from the output of first flip-flop to clock of the next flip-flop for down-counting.

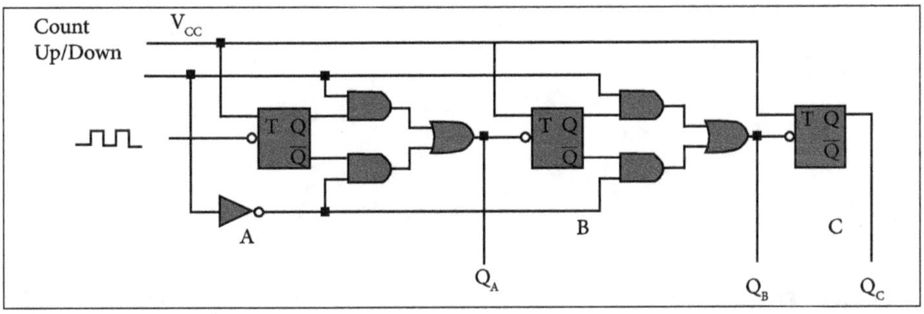

Up-down counter.

As we know that a flip-flop can hold single bit. Therefore, for 3 bit operation, it requires three flip-flops. An inverter has been inserted in between the count-up control line and the count-down control line to ensure that the count-up and count-down cannot be simultaneously in the HIGH state.

When the count-up/down line is held HIGH, lower AND gates will be disabled and their outputs will be zero. Hence, they will not affect the outputs of the OR gates. At the same time, the upper AND gates will be enabled. Hence, QA will pass through the OR gate and into the clock input of the B flip-flop. Similarly, QB will be gated into the clock input of the C flip-flop. Thus, as the input pulses are applied, the counter will count up and follow the natural binary counting sequence from 000 to 111.

Similarly, with the count-up/down line being logic 0, the upper AND gates will become disabled and the lower AND gates are enabled, allowing Q′A and Q′B to pass through the clock inputs of the following flip-flops. Therefore, in this condition, the counter will count in down mode as the input pulses are applied.

2.5.2 Design of MOD-N Counters

Modulus N Synchronous Counters

Designing synchronous MOD 50 counter:

If the number of flip-flops required is n.

Then,

$$2^n = 50 \text{ i.e., } n = 6$$

Since,

$$2^6=64$$

Designing synchronous MOD 25 counter:

If the number of flip - flops required is n, then

$$2^n = 25$$

i.e.,

$$n = 5$$

Since,

$$2^5 = 32$$

Modulo-6 Counter

Let E → Counting enabled.

Present State				Next State			Stable	Output
E	X_3	X_2	X_1	X^+_3	X^+_2	X^+_1	Yes/No	Z
0	0	0	0	0	0	1	Yes	0
0	0	0	1	0	0	1	Yes	0
0	0	1	0	0	1	0	Yes	0
0	0	1	1	0	1	1	Yes	0
0	1	0	0	1	0	1	Yes	0
0	1	0	1	1	0	1	Yes	0
0	1	1	0	1	1	0	Yes	0
0	1	1	1	0	0	0	Yes	0
1	0	0	0	0	0	1	No	0
1	0	0	1	0	1	0	No	0

1	0	1	0	0	1	1	No	0
1	0	1	1	1	0	0	No	0
1	1	0	0	1	0	1	No	0
1	1	0	1	1	1	0	No	0
1	1	1	0	0	0	0	No	1
1	1	1	1	0	0	0	No	0

K-Map Representation

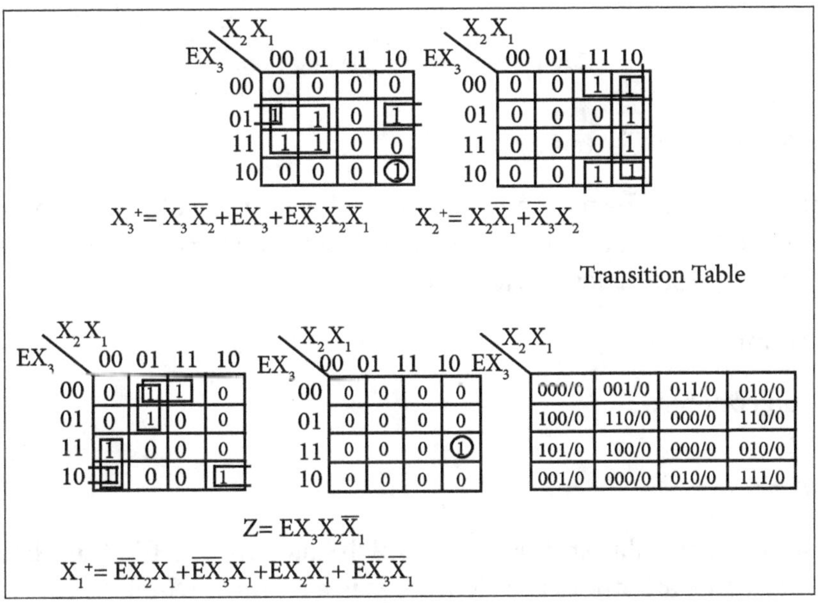

Mod 7 Counter Using D Flip-Flop

A modulo - 7 ripple counter is constructed using D flip flops without the NAND gate. This counter functions as a Modulo - 8 binary counter. The presence of NAND gate alters the sequence as follows:

- The NAND gate output is connected to CLEAR (CLR) inputs of all flip-flops. As long as NAND gate output is HIGH, It will have no effect on counter. When the NAND gate output goes low it will clear all flip-flops and the counter immediately goes to 000 state.
- The outputs of the counter QA, QB, Qc are given as inputs to the NAND gate. The output of NAND gate goes low whenever QA = QB = Qc = 1. This condition will occur when the counter goes from 110 state to 111 state. The low at the NAND gate output will clear the counter to the 000 state.
- Therefore, the counting sequence is → 001 → 010 → 011 → 100 → 101 → 110 → 000 →

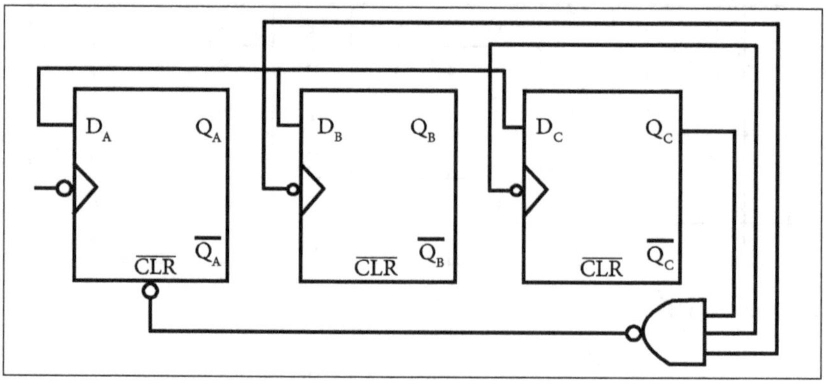

Mod 7 counter using D Flip-flop.

2.5.3 Shift Register Counter

The shift register can also be used as a counter. The shift register with the serial output connection back to the serial input is known as Shift register counter. There are 2 types of shift Register counters. They are as follows:

- Ring counter.
- Johnson counter.

Ring Counter

The ring counter is a circular shift register which has only one flip flop that is being set at any particular time, all others are cleared. It is a counter which counts up and when it reaches the last number which is designed to count up, it will reset itself back to the first number.

Example: The ring counter which is designed using 3 JK Flip Flops may count starting from 001 to 010, 010 to 100 and back to 001. It will repeat itself in a 'Ring'. Therefore, it is called as a ring. counter.

Johnson Counter

A Johnson counter is a K-bit switch-tail ring counter with the 2k decoding gates for providing the outputs for the 2k timing signals. The Johnson counter is said to be a special case of shift register where, the output from the last stage is being inverted and fed back as input to the first stage.

The pattern of bits is equal in length to the shift register so it circulates indefinitely. These counters are sometimes also called as "walking ring" counters. It is used in the specialist applications including those which are similar to the decade counter, digital to analogue conversion, etc.

2.5.4 Decade Counters

A synchronous decade counter counts from zero to nine and repeat the sequence. Since there are ten states, four JK flip-flops are required. The truth tables of present and next state for the decade counter are shown in figure.

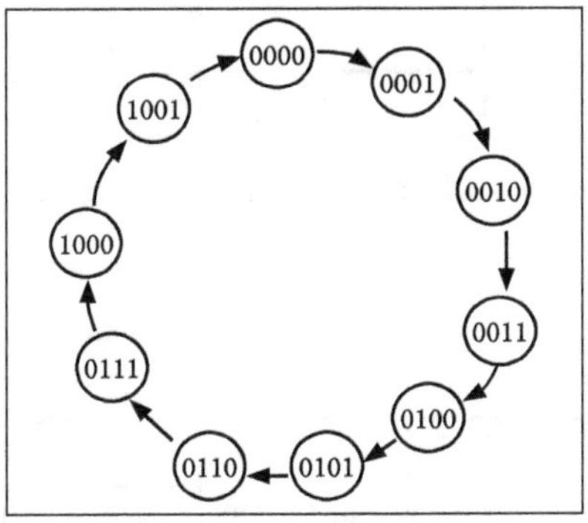

State diagram of synchronous decade counter.

Truth table and state table of synchronous decade counter:

Present state				**Next state**				**Output**							
Q_3	Q_2	Q_1	Q_0	Q_3	Q_2	Q_1	Q_0	J_3	K_3	J_2	K_2	J_1	K_1	J_0	K_0
0	0	0	0	0	0	0	1	0	X	0	X	0	X	1	X
0	0	0	1	0	0	1	0	0	X	0	X	1	X	X	1
0	0	1	0	0	0	1	1	0	X	0	X	X	0	1	X
0	0	1	1	0	1	0	0	0	X	1	X	X	1	X	1
0	1	0	0	0	1	0	1	0	X	X	0	0	X	1	X
0	1	0	1	0	1	1	0	0	X	X	0	1	X	X	1
0	1	1	0	0	1	1	1	0	X	X	0	X	0	1	X
0	1	1	1	1	0	0	0	1	X	X	1	X	1	X	1
1	0	0	0	1	0	0	1	X	0	0	X	0	X	1	X
1	0	0	1	0	0	0	0	X	1	0	X	0	X	X	1

Karnaugh maps of J_0 and K_0:

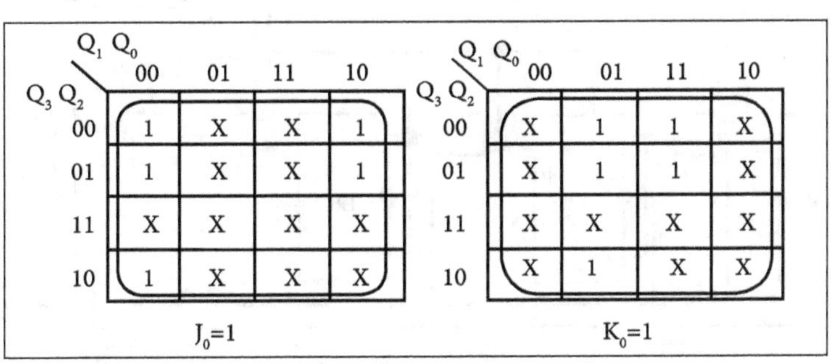

Karnaugh maps of J_1 and K_1:

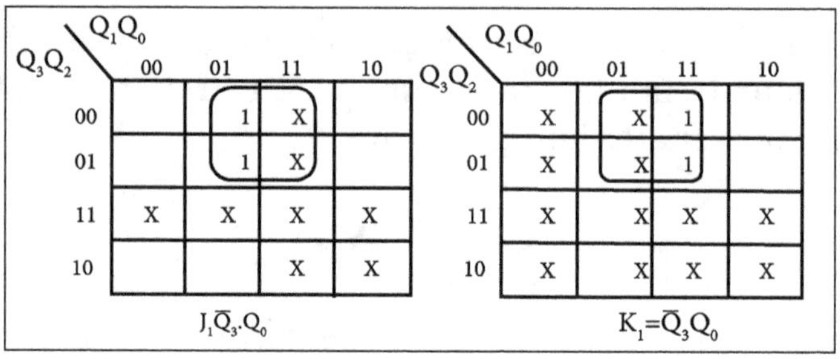

Karnaugh maps of J_2 and K_2:

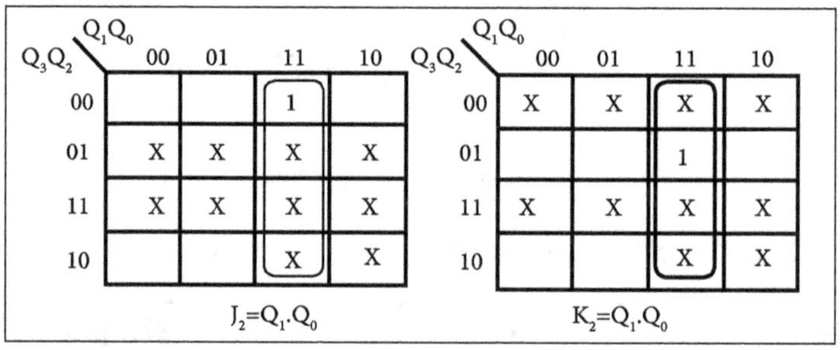

Karnaugh maps of J_3 and K_3:

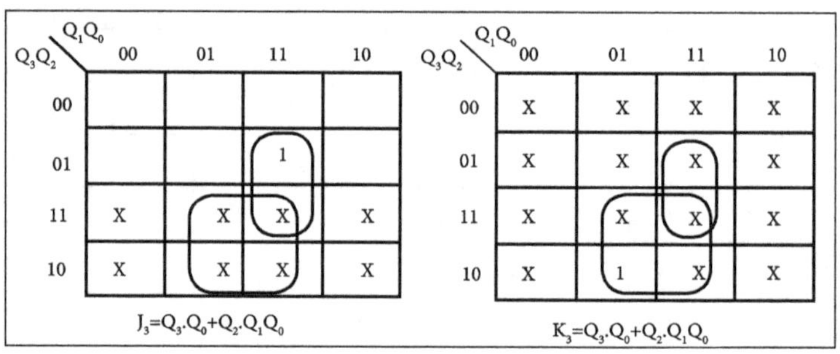

Circuit Implementation

Based on the results obtained from the Karnaugh maps, the circuit design of synchronous decade counter is shown in the below figure.

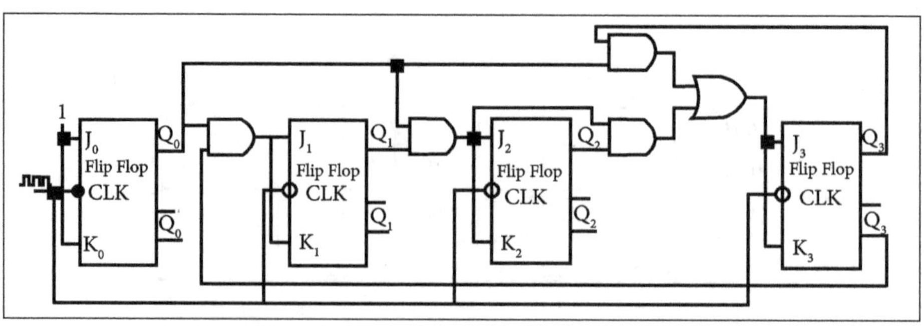

Design of Synchronous decade counter using JK flip-flop.

3

Shift Registers, Memory, Programmable Logic and IC Logic Families

3.1 Shift Registers

The register which are capable of shifting its binary information in one or both the directions from a state to a state within the register or into or out of the register upon the application of clock pulses is known as a shift register. This register is capable of shifting the data, one bit at a time. It is used for converting between the serial and parallel interfaces.

It is useful for as many as circuits to work on groups of bits in parallel, but comparatively the serial interfaces are simpler to construct. The shift registers are used as simple delay circuits and it can also be used as pulse extenders. The serial shift register consists of a chain of flip-flops which is connected in cascade with the output of one flip-flop being connected to the input of its neighbor. The basic operation of the shift register is synchronous.

In shift registers, every flip-flop is connected to a common clock. The D flip-flops forms the simplest type of shift-registers. This basic data movements is possible only within a four-bit shift register which is shown in figure.

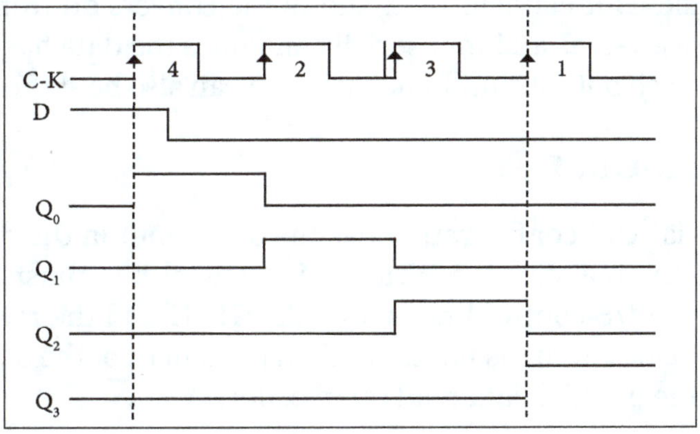

Timing diagram of shift register.

The different types of shift registers are:

- Serial In Serial Out (SISO).

- Serial In Parallel Out (SIPO).
- Parallel In Serial Out (PISO).
- Parallel In Parallel Out (PIPO).

Serial-In Serial-Out

Destructive Readouts

In destructive readout, every datum once lost has been shifted out of the right-most bit. It is the simplest kind of shift register. A data string is presented at 'Data In' and it is shifted in right one at stage where each time the 'Data Advance' is brought high. At every advance, the bit on the far left which is 'Data In' that is shifted into the first flip-flop's output. The bit on the far right is 'Data Out' that is shifted out and lost.

Non-Destructive Readout

The non-destructive readout can be achieved, when another input line is added known as the Read/Write Control. If it is high that is write, then the shift register behaves as normal by advancing the input data one place for each clock cycle and the data can be lost from the end of the register. Since, when the R/W control is set low that is Read, any data can be shifted out of the register at the right becomes the next input at the left and it is kept in the system. Hence, as long as the R/W control is set low, none of the data can be lost from the system.

Serial-In Parallel-Out

The Serial-In, Parallel-Out configuration allows the conversion from serial to parallel format. The data are given as an input serially and once the data has been input, it may be either read off at each output simultaneously or it can also be shifted out and replaced.

Parallel-In Serial-Out

The Parallel-In, Serial-Out configuration has the data input in the parallel format. For writing the data to the register, the Write/Shift control line must be held LOW. For shifting the data, the W/S control line is brought HIGH and the registers are clocked. Until the number of clock cycles is not more than the length of the data-string, the Data Output, Q shall be the parallel data read off in order.

Parallel-In Parallel-Out

In this configuration, the shift register takes the data from the parallel inputs (D_0-D_3)

and shifts it to the corresponding output (Q_0–Q_3) when the registers are clocked. This can be used as a kind of 'history' for retaining the old information as the input in another part of the system as long as it is ready for new information, where upon the registers are clocked and the new data are 'let through'.

Bidirectional Shift Register

The bidirectional shift register allows the shifting of data either to the left or to the right side. This can be implemented by using the logic gate circuitry which enables the transfer of data from one stage to the next stage where either to the right or to the left, depending on the level of a control line. The figure below shows a four-bit bidirectional register.

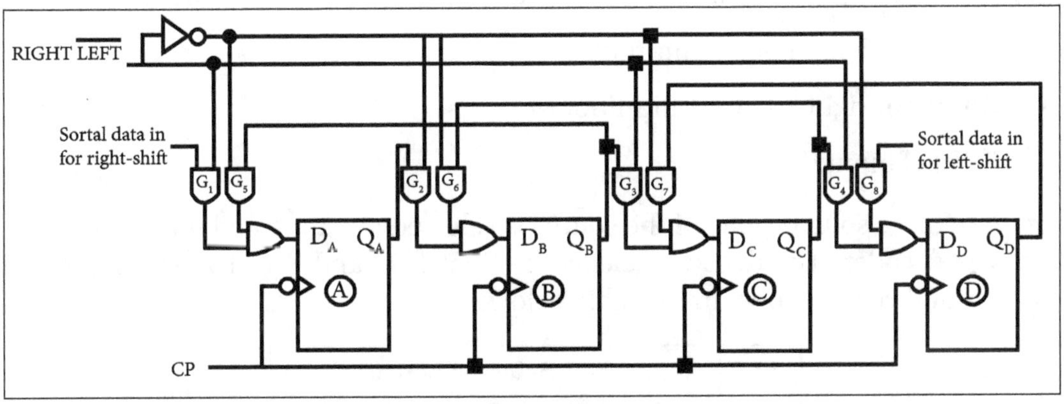

4-bit bi-directional shift register.

The RIGHT or LEFT is the control input signal which allows the data shifting either towards right or towards left. If high on this line, it enables the shifting of data towards the right and a low enables it towards the left. When the RIGHT or LEFT signal is high, the gates G_1, G_2, G_3, G_4 are enabled. The state of the Q output of each flip-flop is passed through the D input of the following flip-flop.

During the arrival of clock pulse, the data are shifted from one place to the right. If the RIGHT or LEFT signal is low, the gates G_5, G_6, G_7, G_8 are enabled. The Q output of each flip-flop is passed through the D input of the preceding flip-flop. During the arrival of clock pulse, the data are shifted from one place to the left.

Bidirectional Shift Register with the Parallel Load

We can see that shift register can be used for converting the serial data into parallel data and vice-versa. While the parallel load capability is added to the shift register, the data entered in parallel can be taken out in serial way by shifting the data stored in the register. This register is known as the Bidirectional Shift Register with parallel load. Figure below shows the bidirectional shift register with parallel load.

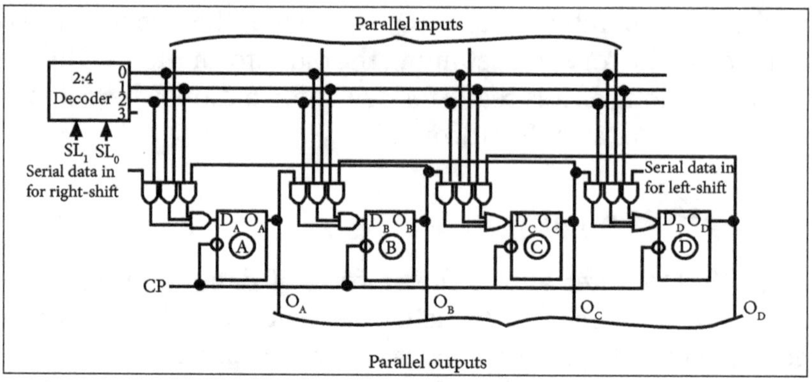

4-bit bi-directional shift register with parallel load.

As shown in the Figure (b), the D input of each flip-flop has three sources:

- Output of left adjacent flip-flop.
- Output of right adjacent flip-flop.
- Parallel input.

From these three sources, one of the source is selected at a time and it is done with the help of the decoder. The decoder selects the lines (SL_1 and SL_0) by selecting the one source out of three as shown in the table below:

SL_1	SL_0	Selected Source
0	0	Parallel input
0	1	Output of right adjacent FF
1	0	Output of left adjacent FF
1	1	-

If the select lines are 00 (i.e., SL_1 = 0 and SL_0 = 0), the data from the parallel inputs is loaded into the 4-bit register. If the select lines are 01 (i.e., SL_1 = 0 and SL_0 = 1), the data within the register is shifted to 1-bit left. If the select lines are 10 (i.e., SL_1 = 1 and SL_0 = 0), the data within the register is shifted to 1-bit right.

3.2 Memory and Programmable Logic

Memory Device

The device to which binary information is transferred for storage and from which information is available for processing as needed is termed as memory devices.

Memory unit is a collection of cells capable of storing a large quantity of binary information.

In digital systems, there are two types of memories such as:

- RAM
- ROM

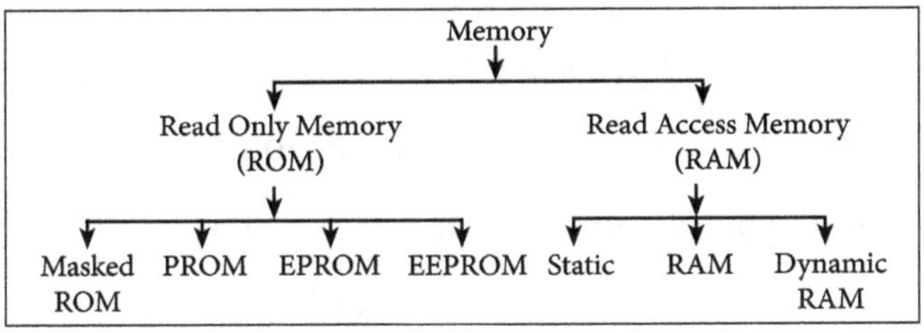

Memory classification.

3.2.1 Memory Decoding

Decoder Operation

- 16 words of 8 bits each.
- A memory with 16 words needs 4 address lines.
- So, 4 x16 decoder is used.
- If memory enable = 0, No memory word is selected.
- If memory enable = 1, one of the 16 word is selected.
- Read/write determines the operation.

Write Operation

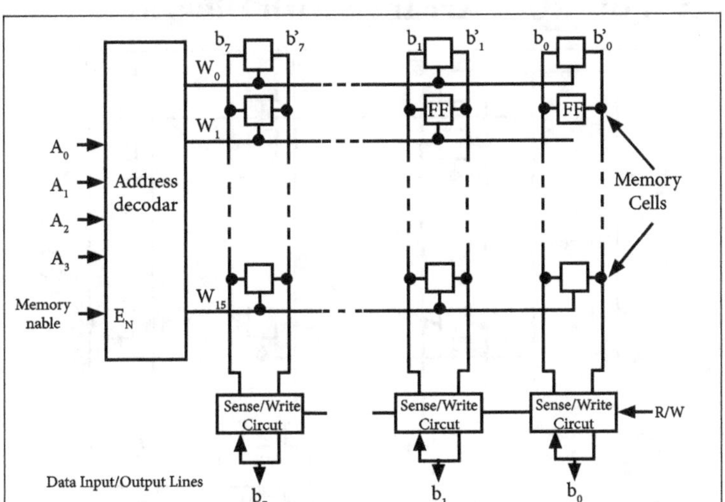

Memory decoding write operation.

- Data is transferred into eight memory cells.

- If a memory cell is not selected, the previous value remain unchanged.

Coincident Decoding

- k input and 2k output.
- 2k and gates are needed with k input to each gate.
- So 2 decoders are used to reduce the no. of inputs.
- So k/2 inputs to each decoder instead of k inputs.
- Instead of 10 x 1024, we use 5 x 32 decoders.

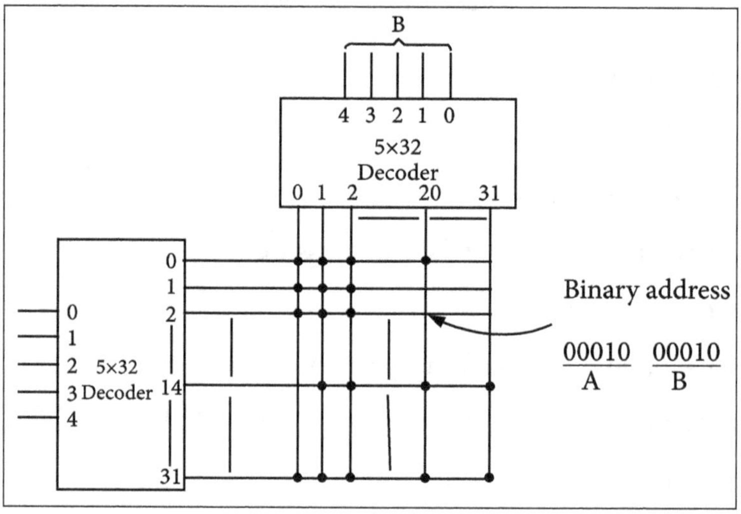

Coincident Decoding.

Concept of 16 X 8 Bit ROM Arrange with Diagram

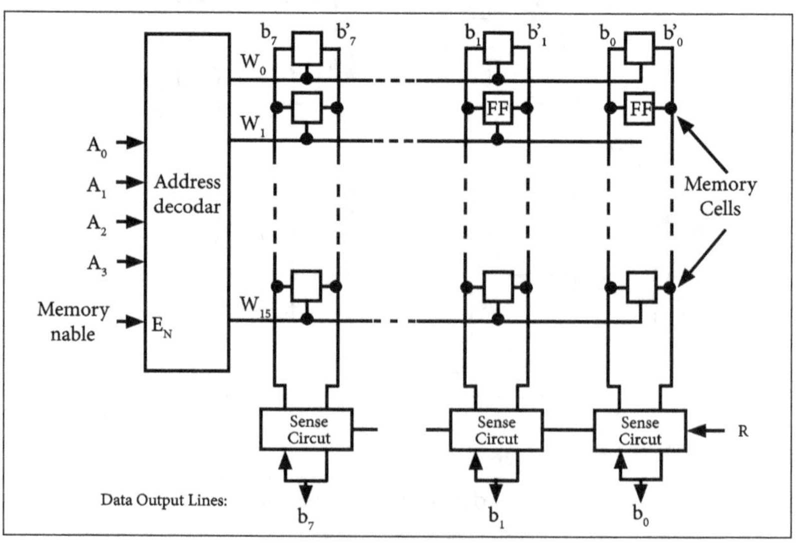

16 X 8 Bit ROM.

Memory Expansion

2 methods:

- Expanding word size.
- Expanding memory capacity.

Expanding Memory Capacity

- By connecting 2 or more ICs together
- Data bus: In Series
- Address bus: In parallel
- Chip select: common to both memory ICs

Design of 1 K X 8 RAM using two 1 K X 4 ICs

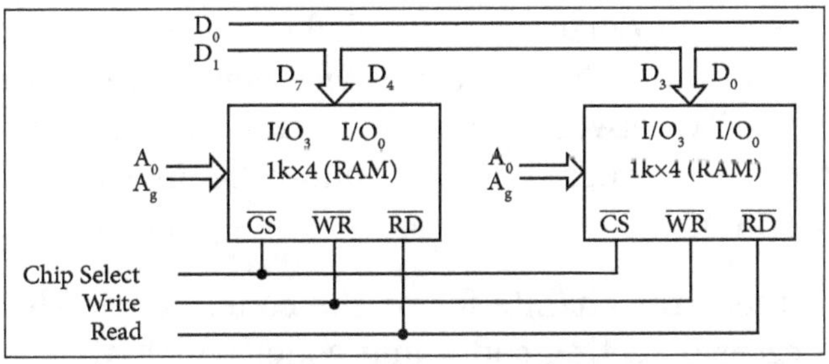

1K X 8 RAM using two 1 K X 4 ICs.

- By connecting 2 or more ICs in parallel. i.e., the address & data bus are connected in parallel.
- Chip select is separate to each cell (generated by address decoder).
- Capacity is limited by the address bus width.

Design of 16 K X 8 RAM using Four 4 K X 8 ICs

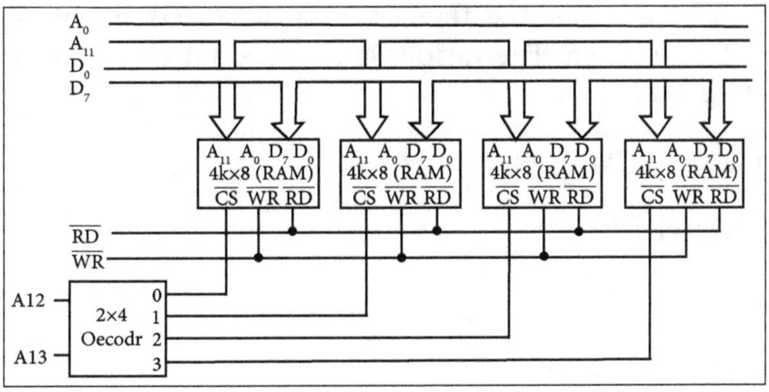

16K X 8 RAM using four 4 K X 8 ICs.

Limitations for Memory Expansion

- Memory devices: Processors accessed using Address, data & Control bus.
- Each Processor has limited no. of address lines & data lines.

Example: Suppose a processor has 24 address lines & 16 data lines, we can expand memory word size up to 16. Memory capacity up to 224 = 16Mbytes

3.2.2 Error Detection and Correction

Error Detecting Codes

When the digital information in the binary form is transmitted from one circuit or system to the circuit or system, an error may occur. To maintain the data integrity between transmitter and receiver, the extra bit or more than one bit are added in the data. The data along with the extra bit/bits forms the code.

Codes which allow only error detection are called error detecting codes. This means a signal corresponding to 0 may change to 1 or vice-versa due to presence of noise. To maintain the data integrity between transmitter and receiver, extra bit or more than one bit are added in the data. These extra bits allow the detection and sometimes, correction of error in the data.

The data along with the extra bit/bits forms the code. Codes which allow only error detection are called error detecting codes and codes which allow error detection and correction are called error detecting and correcting codes.

The most common cause for errors are that the noise creep into the bit stream during the course of transmission from transmitter to the receiver. If these errors are not detected and corrected, the result could be worst as the digital systems are too much sensitive to errors and will malfunction due to the slightest of errors in transmitted codes.

The various methods of error detection and correction such as addition of extra bits that is called check bits, sometimes they are also called redundant bits as they don't have any information in them. The various codes used for error detection and correction code in digital system are:

- Two-dimensional Parity check.
- Simple Parity check.
- Checksum.
- Cyclic redundancy check.

Error Correcting Codes

The method that we have discussed so far can detect errors but it can't correct them. Error Correction can be handled in 2 ways:

- Once when an error is discovered, the receiver might have the sender to retransmit the entire data unit. This is termed as backward error correction.
- The receiver can use an error-correcting code which automatically corrects certain errors. This is known as forward error correction.

In the theory, it is possible to correct any number of errors automatically. The error-correcting codes are more sophisticated than the error detecting codes and need more redundant bits. The number of bits required to correct the multiple-bit or burst error is so high that in most of the cases it is inefficient to do so. For this same reason, most error correction is limited to one, 2 or at the most three-bit errors.

Single-Bit Error Correction

The concept of error-correction can be easily understood by examining the simplest case of single-bit errors. A single-bit error shall be detected by addition of a parity bit (VRC) with the data which required to be send. A single additional bit can detect error, but it is not enough to correct that error too. For collecting an error one has to know the exact position of error, i.e. exactly which bit is in error.

For example, to correct the single-bit error in an ASCII character, the error correction must determine which one of the seven bits is in error. To this we have to add some additional redundant bits. For calculating the numbers of redundant bits (r) required to correct (d) data bits, let us determine the relationship between the two. Therefore, we have (d + r) as the total number of bits which are to be transmitted, then r should be able to indicate at least d + r + 1 different values.

Of these, one value means no error and remaining d + r values indicate error location of the error in each of d + r locations. So. d + r + l states must be distinguishable by r bits and r bits can indicates 2r states. Thus, 2r must be greater than d + r + 1, i.e. 2r > d + r + 1. The value of r must be determined by putting in the value of d in the relation.

Example: If d is 7, the smallest value of r that satisfies the above relation is 4. Thus, the total bits which are to be transmitted is 11 bits (d + r =7 + 4 =11).

Parity Checking

Parity Code is the simplest technique which is used for detecting and correcting errors. The MSB of an 8-bits word is used as the parity bit and the remaining 7 bits are used for data or message bits. Parity bit is added to the transmitted string of bits during the transmission from the transmitters to detect any error in the data when they are received at the receiver end. It is an extra bit added to the string of data. The parity of 8-bits transmitted word can be either even parity or odd parity.

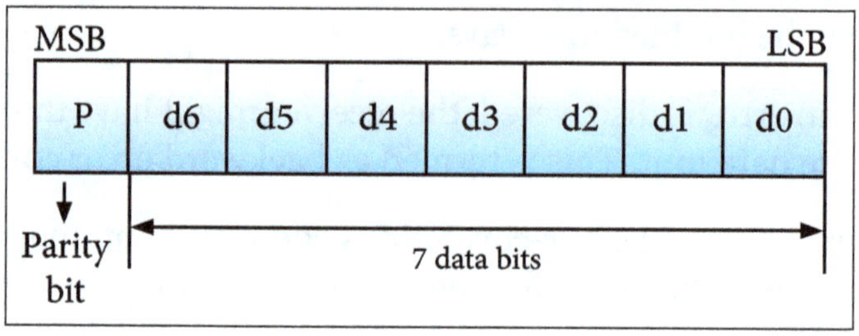

Parity Code.

Parity checking at the receiver can detect the presence of an error if the parity of receiver signal is different from the expected parity. That means, if it is known that the parity of the transmitted signal is always even and if the received signal has an odd parity, then the receiver can conclude that the received signal is not correct. When an error is detected, then the receiver will ignore the received byte and request for the retransmission of the same byte to the transmitter.

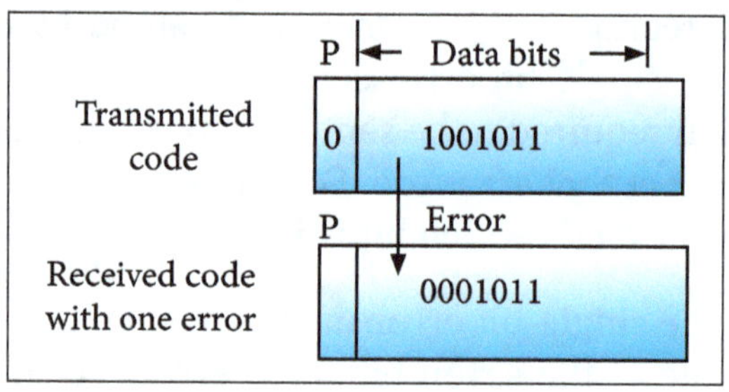

Even Parity

Even parity means the number of 1's in the given word including the parity bit should be even. When we get an even parity, after adding that extra bit the total number of 1's in the string of the data is even.

Odd Parity

Odd parity is defined as the number of 1's in the given word including that the parity bit should be odd. When we get an odd parity, after adding that extra bit into the data string the total number of 1's in the data shall be odd. We can understand it with an example, when we have an eight bit ASCII code – 01000001.

Now if the added bit is 0, then the number will become 001000001. In odd parity, the total number of 1's in the number is even so we get an even parity. Repeatedly if we add 1 to the number the number will become 101000001. The number of 1's is 3 which is odd so we have got an odd parity. Then, normally even parity is used and it has almost become a convention.

Now the parity checks are capable of detecting the single bit error. Since it fails, there are two changes in the data, which is the biggest drawback of this system. Hence, there are several other codes to detect and correct more than one bit errors.

Use of Parity Bit

- The parity bit can be set to 0 and 1 where it depends on the type of the parity required.
- For even parity, this bit is set to 1 or 0 whereas the no. of the 1 bits in the entire word is even which is shown in figure (a).
- For odd parity, this bit is set to 1 or 0 whereas the no. of the 1 bits in the entire word is odd which is shown in figure (b).

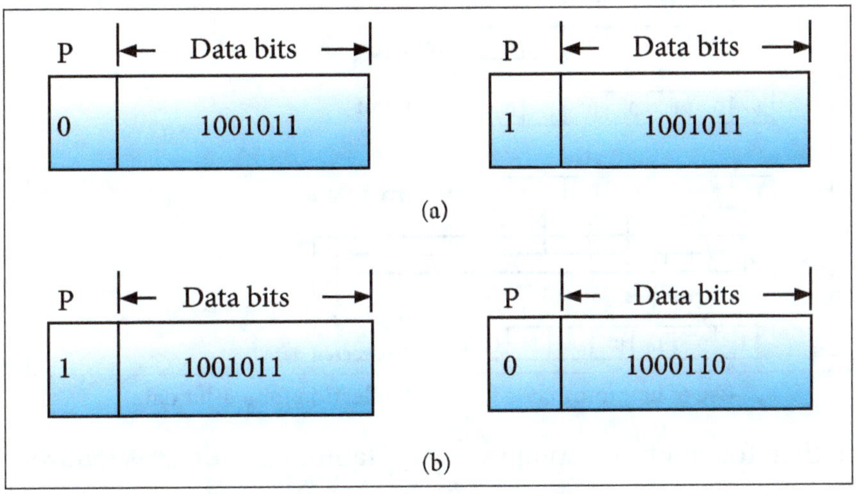

(a) Even parity (b) Odd parity.

Hamming Code

Let us examine the manipulation with these bits to discover which bit is in error. The method developed by R.W. Hamming gives a practical solution. The solution or coding scheme, he developed is commonly called as Hamming Code. The hamming code can be applied to data units of any length and which uses the relationship between the data bits and the redundant bits.

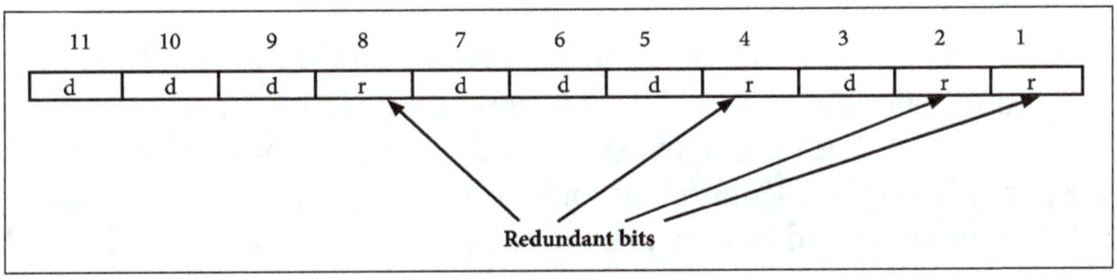

Positions of redundancy bits in hamming code.

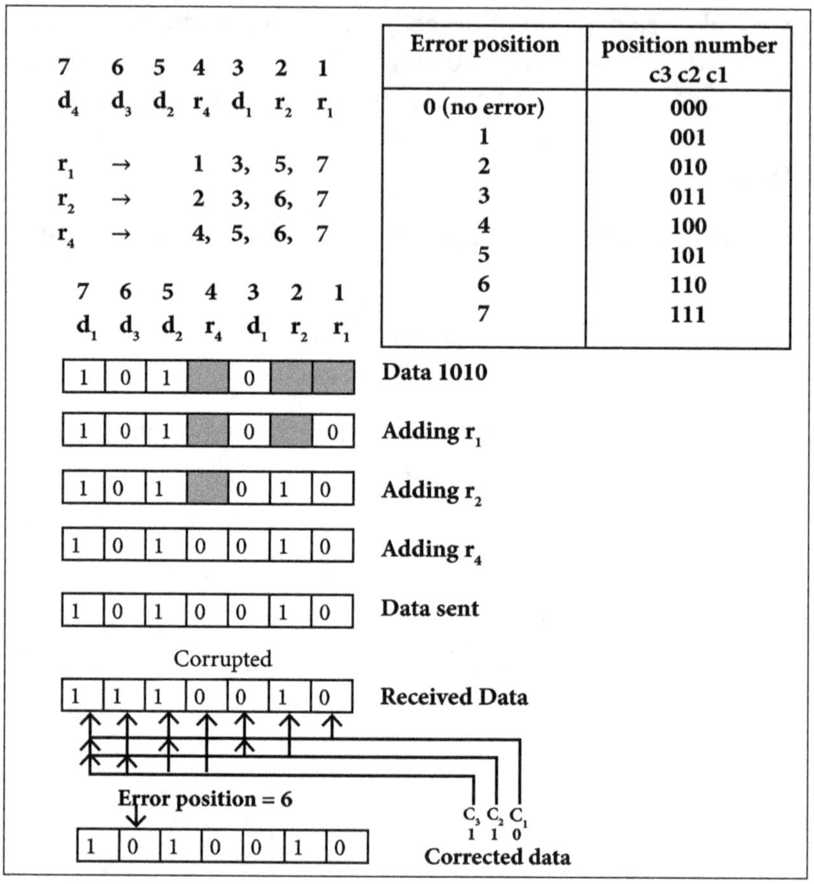

Error position	position number c3 c2 c1
0 (no error)	000
1	001
2	010
3	011
4	100
5	101
6	110
7	111

Use of hamming code for error collection for a 4-bit data.

The basic method for error detection by using Hamming code is as follows:

- The location of each of the (m + K) digits is assigned a decimal value.
- To each group of m information bits k parity bits are added to form (m + k) bit code as represented in figure.
- At the receiving end, the parity bits can be recalculated. The decimal value of the k parity bits provides the bit-position in error, if any.
- The k parity bits are placed in positions 1, 2, 2K-1 and K, parity checks are performed on selected digits of each code word.

The above figure represents how the hamming code is used for the correction of 4-bit numbers ($d_4d_3d_2d_1$) with the help of three redundant bits ($r_3r_2r_1$). For example, consider data 1010. First r_1 (0) is determined considering the parity of the bit positions 1, 3, 5 and 7. Then the parity bits r_2 is calculated by considering bit positions 2, 3, 6 and 7. Finally, the parity bits r_4 is calculated by considering the bit positions 4, 5, 6 and 7 as shown.

If any corruption occurs in any of the transmitted code 1010010, the bit position in the error can be found out by calculating the $r_3r_2r_1$ at the receiving end. For example, if the received code word is 1110010, the recalculated value of the $r_3r_2r_1$ is 110 that indicates that bit position in error is 6, the decimal value of 110:

- Parity Generator: It is the circuit that generates the parity bit in the transmitter.
- Parity Checker: It is the circuit that checks the parity in the receiver.

3.2.3 RAM and ROMs

RAM

A memory unit is a set of storage cells together with associated circuits required to transfer information into and out of a device. The time it takes to transfer information to or from any desired random location is always the same. Hence, the name random access memory, abbreviated as RAM.

A memory unit keeps binary information in groups of bits called words. A memory word is a group of 1's and 0's and may represent a number, an instruction, one or more alphanumeric characters or any other binary-coded information. A group of 8 bits is called a byte. Most computer memories use words that are multiples of 8 bits in length.

Each word in memory is assigned an identification number called an address starting from 0 up to 2^k -1, where k is the number of address lines. Consider for example, a memory unit with a capacity of 1K words of 16 bits each. Since 1K = 1024 = 2^{10} and 16 bits constitute 2 bytes, we can say that the memory can accommodate 2048 = 2K bytes.

Memory address		
Binary	Decimal	Memory content
0000000000	0	1011010101011101
0000000001	1	1010101110001001
0000000010	2	0000110101000110
	• • • •	• • • •
1111111101	1021	1001110100010100
1111111110	1022	0000110100011110
1111111111	1023	1101111000100101

Contents of a 1024 × 16 memory.

In general, RAM refers to random access memory. The term RAM has also come to mean a memory which can be both easily written to and read from. There are two main technologies used for RAM:

1. Static RAM

These are arrays of flip-flops. They can be fabricated in ICs as large arrays of thin flip-flops. "SRAM" is intrinsically somewhat faster than dynamic RAM.

2. Dynamic RAM

It uses capacitor arrays. Charge put on a capacitor will produce a HIGH bit if its voltage V = Q/C exceeds the threshold for the logic standard in use. Since the charge will "leak" o through the resistance of the connections in times of order 1msec, the stored information must be continuously refreshed (hence the term "dynamic").

Dynamic RAM can be fabricated with more bits per unit area in an IC than static RAM. Hence, it is usually the technology of choice for most large-scale IC memories.

Read Operation of SRAM

The complementary bit lines are pre charged to an intermediate voltage between VDD and ground. The bit lines are connected to the latches through access transistors when the cell is selected. The differential voltage develops between complementary bit lines. The voltage changes and it should be sufficiently small not to change the latch state during readout. The read operation in an SRAM is nondestructive.

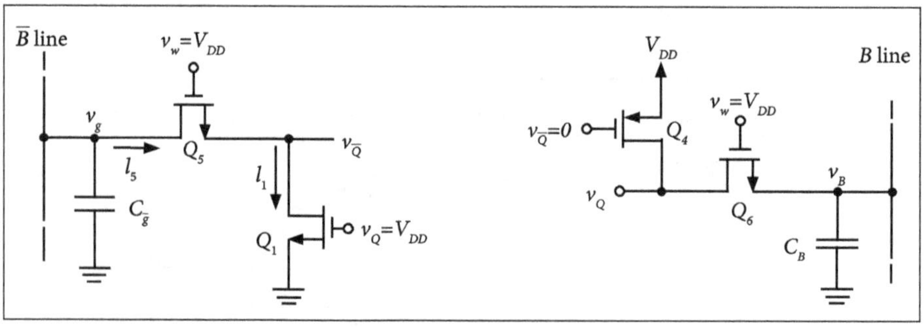

Read operation of SRAM.

Readout delay is determined by the rise time of word line and time needed to develop the required voltage. The dynamic operation can be approximated by treating Q_1 and Q_5 as r_{ON1} and rON5.

Design Constraint for Read Operation

The worst case scenario is to choose a pre charge value of V_{DD}. Q_5 operates in saturation region.

Q_1 operates in triode region,

$$I_5 = \frac{1}{2}\mu_n C_{ax}\left(\frac{W}{L}\right)_5 \left(V_{DD} - V_{in} - V_{\bar{Q}}\right)^2 = I_1 = \mu_n C_{ax}\left(\frac{W}{L}\right)_1 \left[\left(V_{DD} - V_{in}\right)V_{\bar{Q}} - \frac{1}{2}V_{\bar{Q}}^2\right]$$

The design constraint is specified by,

$$V_{\bar{Q}} = \left(V_{DD} - V_{in}\right)\left(1 - \frac{1}{\sqrt{1 + \frac{(W/L)_5}{(W/L)_1}}}\right) \le V_{in} \rightarrow \frac{(W/L)_5}{(W/L)_1} \le \frac{1}{\left(1 - \frac{V_{in}}{V_{DD} - V_{in}}\right)^2} - 1$$

The voltage change in v_Q will be very small. The $\bar{B}$ line is discharged by I_5 at the beginning,

$$\Delta V = I_5 \Delta t / C_{\bar{B}}$$

The time needed to develop a voltage difference ΔV is given by,

$$\Delta t - C_{\bar{B}} \Delta V / I_5$$

The $\bar{B}$ line will be finally discharged to o in steady state. Sense amplifiers are usually used to speed up the read operation.

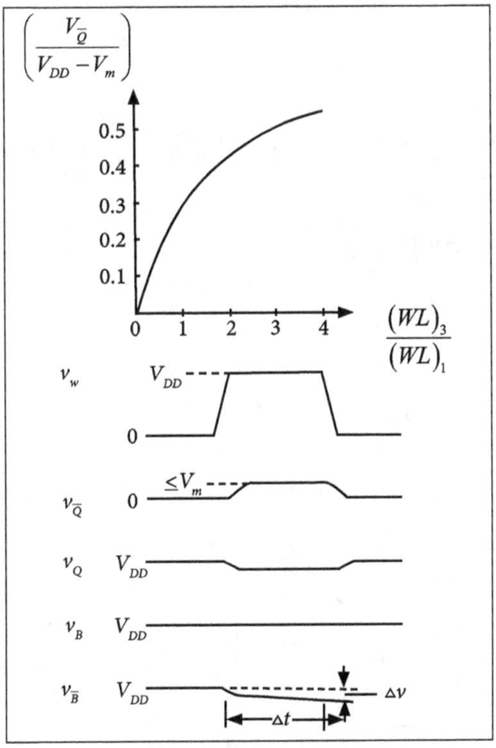

Read Operation of SRAM.

Write Operation of SRAM

The complementary bit-lines are respectively set to V_{DD} and ground before the operation. The cell is selected by the word line and the cell is connected to the bit lines through access transistors. Q_5 is in saturation and the current for charging capacitor equals to $I_5 - I_1$. I_5 decreases (due to reduced VGS5 and body effect) and I_1 increases (due to reduced VGS1). When v_Q and $V_{\bar{Q}}$ reach $V_{DD}/2$, the regenerative feedback initiates and cause the flip-flop to change state.

Write delay is determined by the time for v_Q and $V_{\bar{Q}}$ to reach $V_{DD}/2$ and the delay time of the flip-flop. The charging (discharging) component of write delay is much smaller than the corresponding component in the read operation because only small capacitance C_Q needs to be charged (discharged). The delay time in write operation is dominated by the word-line delay.

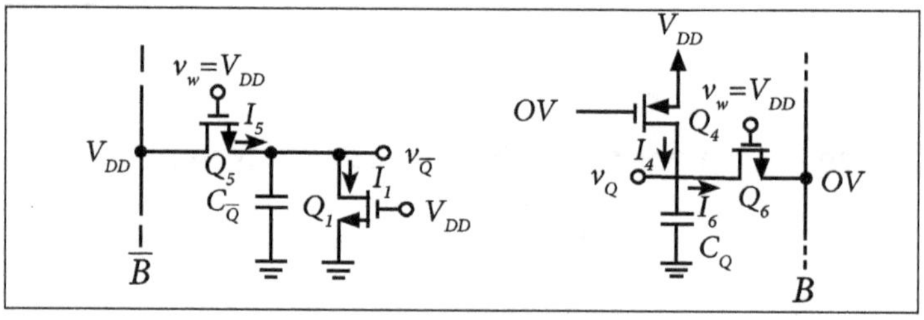

Write operation of SRAM.

Design Constraint for Write Operation

Q_4 operates in saturation region,

$$I_5 = \frac{1}{2}\mu_n C_{ax}\left(\frac{W}{L}\right)_4 \left(V_{DD} - |V_{in}|\right)^2 = I_6 = \mu_n C_{ax}\left(\frac{W}{L}\right)_1 \left[\left(V_{DD} - V_{in}\right)V_{\bar{Q}} - \frac{1}{2}V_{\bar{Q}}^2\right]$$

The design constraint is specified by,

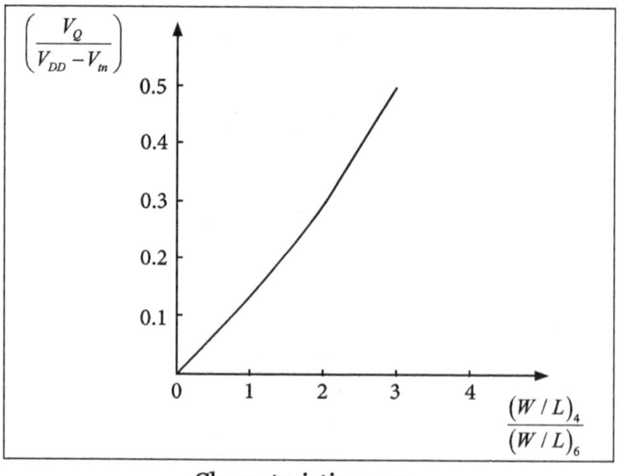

Characteristic curve.

$$V_Q = (V_{DD} - V_{in})\left(1 - \sqrt{\frac{\mu_p (W/L)_4}{\mu_n (W/L)_6}}\right) \le V_{in} \rightarrow \frac{(W/L)_4}{(W/L)_6} \le \frac{\mu_p}{\mu_n}\left[1 - \left(\frac{V_{in}}{V_{DD} - V_{in}}\right)^2\right]$$

Dynamic RAM Cell

Read Operation

The bit lines are pre charged to $V_{DD}/2$. The word line is raised to VDD and all the access transistors in the selected row are conductive. A small voltage difference will be detected in the bit lines. The read operation is destructive,

$$C_s V_{cs} + C_B \frac{V_{DD}}{2} = (C_B + C_s)\left(\frac{V_{DD}}{2} + \Delta V\right)$$

$$\Delta V = \frac{C_s}{C_B + C_s}\left(V_{cs} - \frac{V_{DD}}{2}\right) \approx \frac{C_s}{C_B}\left(V_{cs} - \frac{V_{DD}}{2}\right) \text{ for } C_B >> C_s$$

Write Operation

The data bit to be written is set to V_{DD} or ground and C_S will be charged or discharged to V_{DD}-V or 0V. All other cells in the selected row perform write operation at the same time.

Refresh Operation

Read operation followed by a write operation for stored data.

Timing Diagram

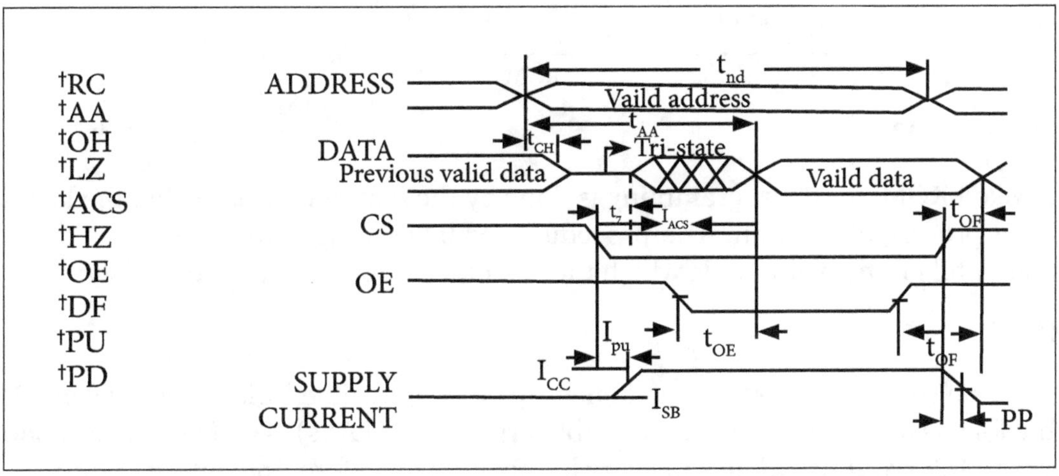

Read cycle.

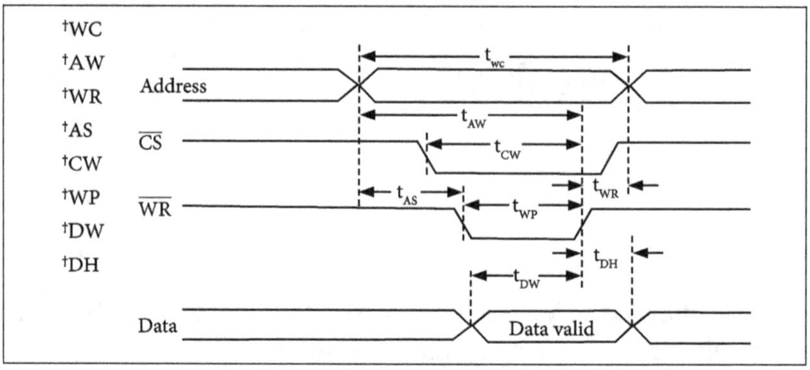

Write cycle.

ROM: Read-Only Memory

Information cannot be easily stored. The bits are initially decoded and are never changed thereafter.

ROM Organization

It is a read only memory. We can't write data in this memory. It is non-volatile memory i.e., it can hold data even if power is turned off. There are four types of ROM such as Masked ROM, PROM, EPROM and EEPROM or E^2PROM.

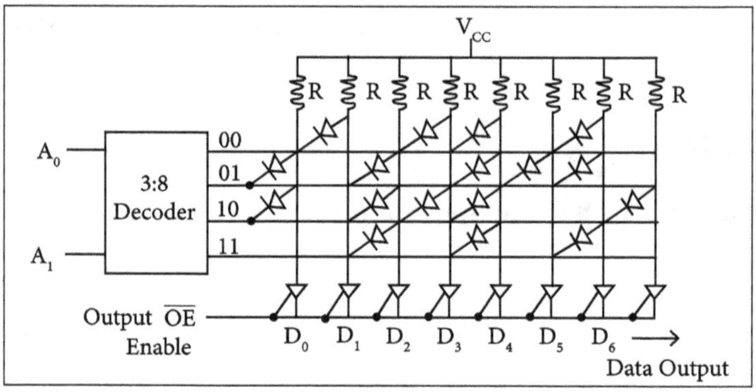

ROM

Masked ROM

In masked ROM, mask programming is done by the manufacturer during the last fabrication process of the unit. The procedure for fabricating a ROM requires that the customer fill out the truth table which the customer wants the ROM to satisfy.

PROM

Bits can be set on a programming bench by burning fusible links or equivalent. This technology is also used for programmable array logic (PALs). The figure below shows four byte PROM. It has diodes in every bit position therefore, the output is initially all 0's.

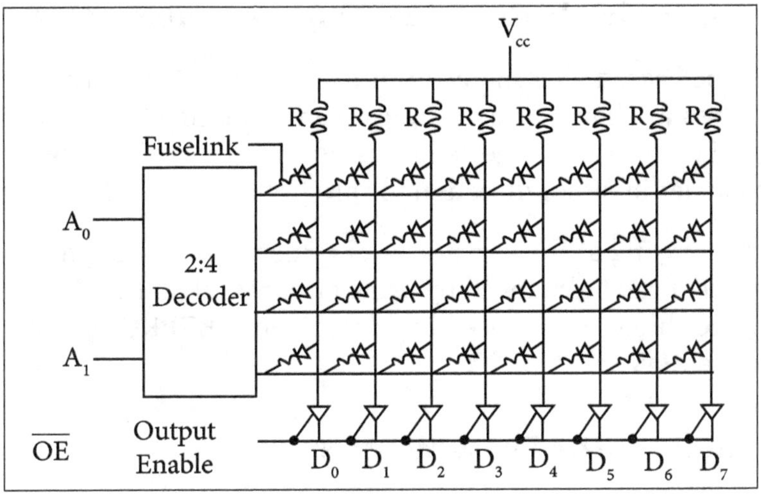

PROM

Each diode, however has a fusible link in series with it. By addressing bit and applying proper current pulse at the corresponding output, we can blow out the fuse, storing logic 1 at that bit position. The PROM programmer selectively burns the fuses according to the bit pattern to be stored. This process is known as burning of PROM. It is one time programmable. Once programmed, the information stored is permanent.

EPROM: Erasable Programmable ROM

It is a type of ROM which can be erased using ultraviolet light. The erasable PROM (EPROM) is also useful for developmental and experimental work because its contents can be programmed with relative ease and at small cost. The principal feature of the EPROM is that its contents can be erased and reprogrammed, enabling the same device to be used repeatedly. The process of programming an EPROM involves the application of special voltage levels to the appropriate chip inputs for a specified amount of time.

The complete programming process can take up several minutes for one EPROM chip. To erase an EPROM, the cells must be exposed to relatively intense UV radiation and it has transparent quartz window. This erasing process typically requires 15 to 30 minutes of exposure to UV rays. Once erased, the EPROM can be reprogrammed. Once an EPROM has been programmed an opaque cable is placed over the window to prevent stray radiation from causing gradual erasure. EPROMs are available in a wide range of capacities and access times. A popular line of commercially available EPROMs is the 27xx series.

Difference between PROM and EPROM

PROM	EPROM
PROM is Programmable Read Only Memory. It consists of a set of fixed AND gates connected to a decoder and a programmable OR array. The programmable OR gates implement the Boolean functions in sum of minterms.	EPROM is Erasable and Programmable Read Only Memory. The ROM can be erased and programmed again and again.

EEPROM: Electronically Erasable Programmable ROM

ROM which can be erased electronically. It also use MOS circuitry similar to that of ROM. Data is stored as charge or no charge on an insulating layer, which is made very thin (< 200 A). Therefore a voltage as low as 20 to 25 V can be used to move charges across the thin barrier in either direction for programming ROM or erasing.

It allows selective erasing at the register level rather than erasing all the information, since the information can be changed by using electrical signals. It has chip erase mode by which the entire chip can be erased in 10 ms. Hence EEPROMs are most expensive and the least dense ROMs.

Electrically Alterable or Erasable Prom

The EAPROM or EEPROM can be programmed (modified) or erased while it is still in the circuit and used like a nonvolatile read/write memory. EAPROM/EEPROM use an electric charge to erase the contents.

Some types of EAPROM/EEPROM are more versatile. Individual cells can be reprogrammed by reversing the voltage used to create a zero. There are some timing constraints that causes more time for erasure or programming than is needed to read data from the part. Some EAPROM/EEPROM have a word or byte erase mode.

3.3 Programmable Logic Array

A programmable logic array (PLA) is a kind of programmable logic device used to implement the combinational logic circuits. The PLA has a set of programmable AND gate planes, which connected to a set of programmable OR gate planes, which can then be conditionally complement an output. This layout permits for a large number of logic functions to be synthesized in the sum of products canonical forms.

PLA's differ from Programmable Array Logic devices (PALs and GALs) in that both the AND and OR gate planes are programmable. The logic gates used can be two level AND-OR, NAND-NAND or NOR-NOR configuration. The PLA consist of two level AND-OR circuit on the same chip. The AND matrix can be used to implement the product terms in the SOP form and the OR matrix is used for implementing the sum of product terms.

Input Buffers

Input Buffer in a PLA is used for avoiding the loading of sources connected at the input.

Block Diagram

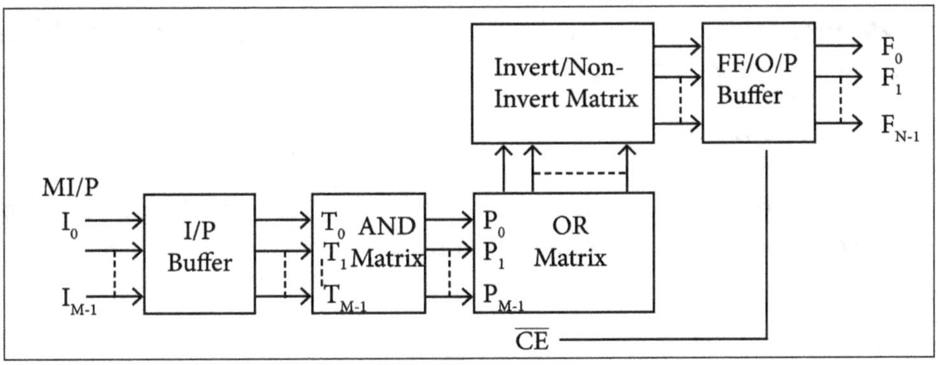

PLA

Internal Construction

In PLA, both the AND and OR gates are programmable. The internal construction of PLA having 3 inputs, 3 product terms and two outputs, is as shown below.

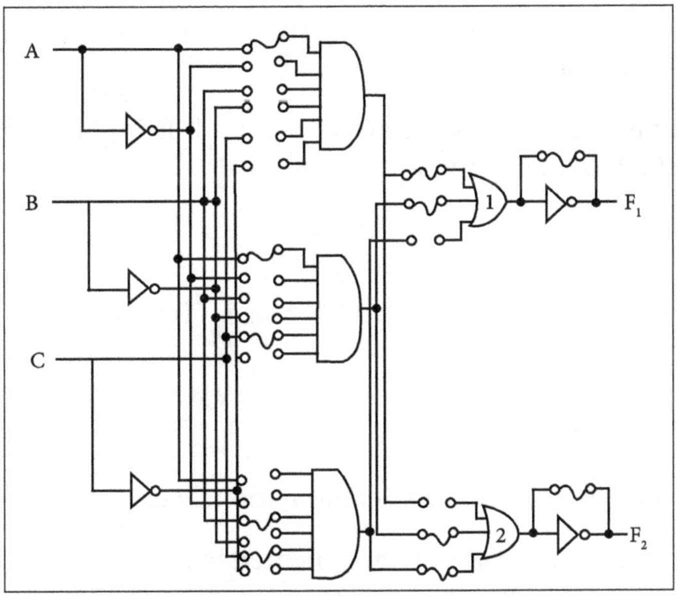

Internal construction of PLA.

The size of PLA is specified by number of inputs, the number of product terms and the number of outputs. A PLA can be mask-programmable or field programmable. PLA consists of inputs, outputs, sum terms and product terms. The product terms constitute a group of AND gate and sum terms constitute a group of OR gates. Fuses are inserted between all inputs and their complement value to each of the AND gates.

Fuses are also provided between the output of AND gates and the inputs of OR gates. The third set of fuses in the output inverters allows the output function to be generated either in AND-OR form or in the AND-OR-INVERT form.

AND Matrix

There are 2M input (from) for each AND gate, P_0 is the output of an AND gate. The Boolean expression for P_0 is given by,

$$P_0 = I_0 - \overline{I}_0 - I_1.\overline{I}_1 ... I_{M-1}\,\overline{I}_{M-1}$$

There is a nichrome fuse link connected in series with each input. The above expression produces $P_0 = 0$ if all the fuse linker are intact.

OR Matrix

P_0, P_1, ..., P_{M-1} are the output of AND matrix. They are used as the inputs to the OR matrix. The OR gate is formed by parallel connected transistors and the load is common and connected in the emitters. S_0, S_1, ... , S_{N-1}, are the output of the OR matrix.

Invert / Non - Invert Matrix

The block following the OR matrix is the inverting/non - inverting matrix which is basically a programmable buffer.

It can invert its input if active low output is required. It input is passed without any inversion if active high output is required.

Output Buffer

The current sourcing capability of the PLX can be increased by using the output buffer.

Advantages of PLA'S

- It is used to implement complex combinational circuit.
- Specific Boolean functions are implemented in sum of products (SOP).

Advantages of PLA Over ROM

- PLA is advantageous for cases where the number of don't care conditions are excessive, as PLA is more economical than ROM.
- A PLA is similar to a ROM in concept, however it does not provide full decoding of the variables and does not generates all the minterms as in the ROM.
- The PLA replaces the decoder by group of AND gates, each of which can be programmed to generate a product term of input variables.

3.3.1 Programmable Array Logic

Programmable Array Logic (PAL) is a family of programmable logic device semiconductors used to implement logic functions in digital circuits introduced by Monolithic Memories, Inc. (MMI) in March 1978. MMI obtained a registered trademark on the term PAL for use in "Programmable Semiconductor Logic Circuits". The trademark is currently held by Lattice Semiconductor.

PAL devices consist of a small PROM (programmable read-only memory) core and the additional output logic used to implement particular desired logic functions with few components. Using specialized machines, PAL devices were "field-programmable". PALs were available in several variants.

"One-time programmable" (OTP) devices cannot be updated and reused after initial programming. UV erasable versions (e.g.: PALCxxxxx, e.g.: PALC22V10) had a quartz window over the chip die and could be erased for re-use with an ultraviolet light source just like an EPROM.

Later versions (PALCExxx e.g.: PALCE22V10) were flash erasable devices.

Comparison between PROM, PLA, PAL

S.No.	PROM	PLA	PAL
1.	AND array is fixed and OR array is programmable.	Both AND & OR arrays are programmable.	AND array is programmable and OR array is fixed.
2.	Cheaper and Simple.	Costliest & complex than PAL & ROMS.	Cheaper and simpler.
3.	All minterms are decoded.	AND array and OR array can be programmed to get desired minterms.	AND array can be programmed to get desired minterms.
4.	Only Boolean functions in canonical SOP form can be implemented using PROM.	Any Boolean function in standard SOP form can be implemented using PLA.	Any Boolean function in standard SOP form can be implemented using PAL.

3.3.2 Sequential Programmable Devices

Digital systems are designed with flip-flops and gates. Since the combinational PLD consists of only gates, it is necessary to include external flip-flops when they are used in the design. Sequential programmable devices include both gates and flip-flops.

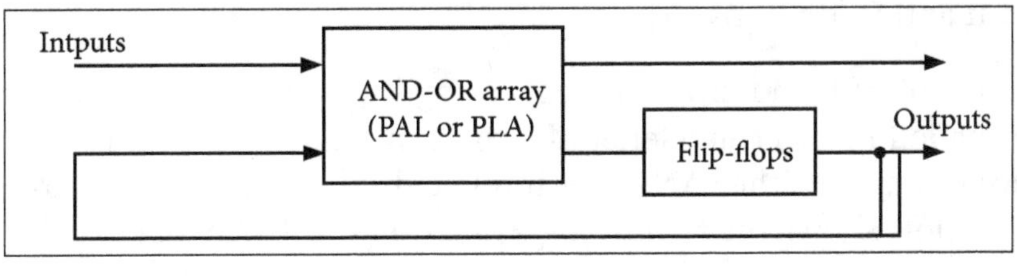

Digital systems are combination of combinational circuits and sequential circuits. A digital circuit without sequential part is not complete. Hence, combinational PLDs is in need of external flip flops. Sequential programmable devices consist of both flip flops and combinational circuits.

Three major types of SPLD are:

- Simple Programmable Logic Devices (SPLD).
    - PROM (Programmable Read Only Memory).
    - Programmable Logic Array (PLA).
    - Programmable Array Logic (PAL).
    - Generic Array Logic (GAL).
- Complex Programmable Logic Device (CPLD).
- Field Programmable Gate Array (FPGA).

1. Sequential Programmable Logic Device (SPLD)

Sequential or Simple PLD (SPLD) consists of flip flops and AND-OR arrays as an integrated circuit. The block diagram of a SPLD is shown in Figure.

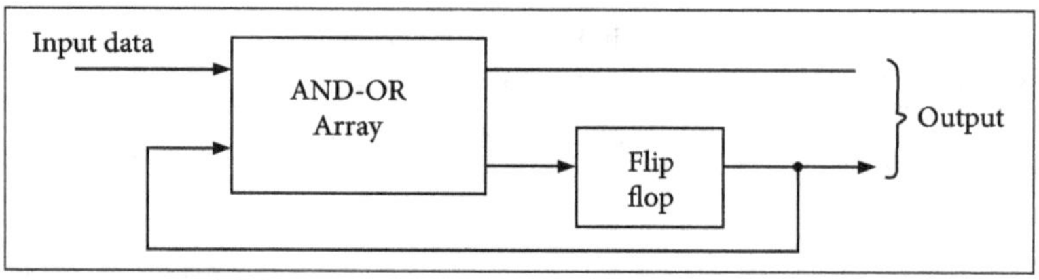

Sequential (or simple) programmable logic device (SPLD).

Cascading of PAL (or PLA) with set of flip flops becomes sequential (simple) programmable logic device. The output of the circuit can be taken either from flip flops or directly from the OR gates. A feedback provision from flip flops to array of AND gates is also there. The flip flops are either D type or JK type.

The first sequential PLD developed is the Field Programmable Logic Sequential (FPLS). In this case, the flip flops are driven by the AND arrays and OR gates. The flip flops itself can be programmed either as JK flip flops or D flip flops. But this FPLS consist of large number of programmable connections. Most of the SPLDs use PAL type combinational circuit and D flip flops.

The combination of PAL and flip flop is called registered PAL. Each group of AND - OR - Flip flops is called macro cell. Generally, the combinational part is sum of products but in practice it may be either AND-OR implementation or NAND-NAND implementation. Figure below shows the typical logic diagram of a basic macro cell.

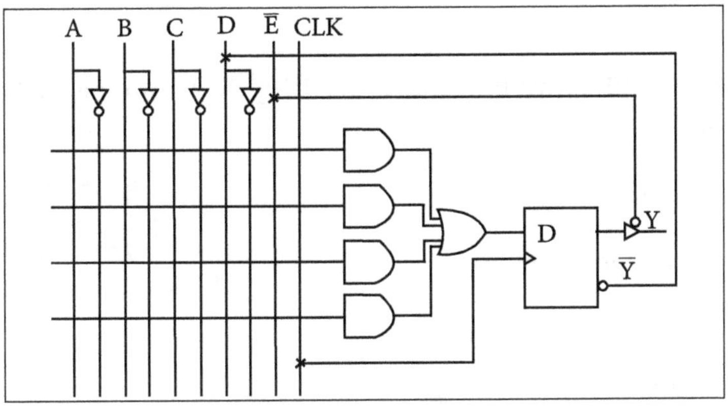

Basic macro cell.

The output is the function of D flip flop driven by external common clock. The output is driven through tristate buffer or inverter with a provision of output enable signal. The feedback of the output to input is also possible. The SPLD ICs are manufactured in such way that each will contain 8 to 10 macro cells. A common clock input for flip flop and output enable signal for buffer are provided.

Similar to programming the AND array in PAL and PLA, the macro cell can be programmed for positive or negative edge trigger, set and reset of flip flops and complement form of outputs. Similar to PLA, SPLD can also use XOR gates to complement the output. Selection of input can also be programmed.

PROM

PROM consists of a set of fixed AND gates connected to a decoder and a programmable OR array. The programmable OR gates implement the Boolean functions in sum of minterms. It is used for code conversions, generating bit patterns for characters and as Look Up Tables (LUT) for arithmetic functions.

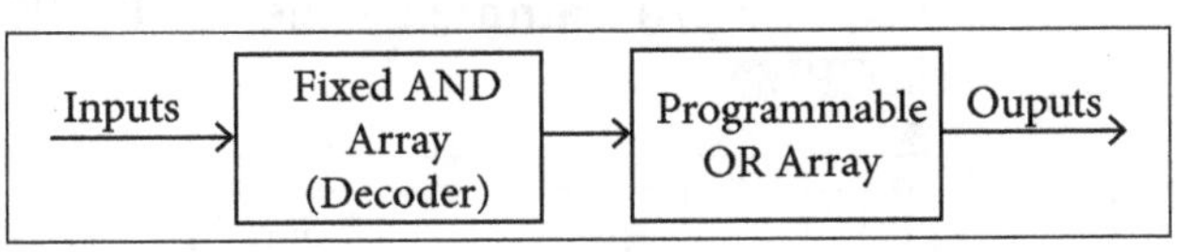

PROM

Eg,

$$F_1(X, Y, Z) = \sum m(1,2,4,7)$$
$$F_2(X, Y, Z) = \sum m(3,5,6,7)$$

PLA

It is a type of fixed architecture logic devices with programmable AND gates followed by programmable OR array. PLA is used to implement a complex combinational circuit.

PAL

It is a type of fixed architecture logic devices with programmable AND array followed by fixed OR gates.

Advantages

- Reduced space requirement.
- Increased speed of switching.
- Higher density.
- Low development cost.

2. Complex Programmable Logic Device (CPLD)

The complete specification of SPLD requires cascading connections of many devices. For such applications, CPLD will play a better role for economical operation. A CPLD is a collection of PLDs on a single IC. The cascading of PLDs is similar to connecting individual PLDs. A typical block diagram of a CPLD is shown in Figure.

The CPLD consists of many PLDs interconnected through programmable switch matrix. The interface between external devices and CPLD made through I/O block through IC pins. Each pin in the IC is internally connected through bidirectional tristate buffers. The direction of bidirectional tristate buffers is programmable and it will be either input buffer mode or output buffer mode.

I/O block directs the input to the respective macro cell by using the switch matrix and directs the output to the respective output buffers. Each PLD consists of 8 to 16 macro cells. One micro cell in a particular PLD can be used by a macro cell in the other PLDs. The flip flop in the macro cell can be programmed either as D, JK or T flip flops. The given specifications are generic CPLD.

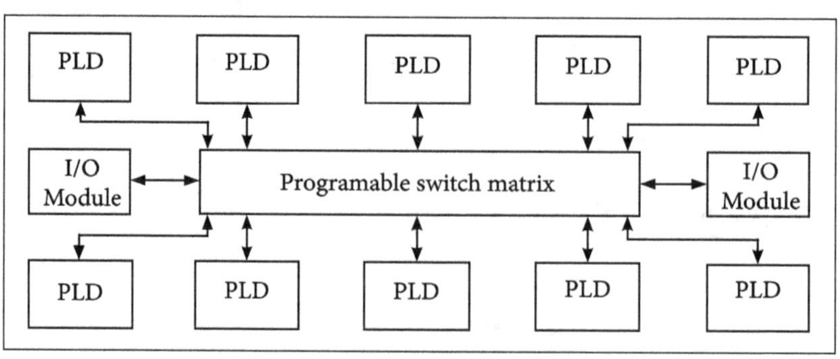

Generic CPLD Configuration.

The way in which individual CPLDs are defined, the nature of macro cells, functions of I/O blocks and cascade connections of CPLDs vary from vendor to vendor. The exact configuration of a specific CPLD can be correctly obtained from the manufacturer's data sheet.

3. Field Programmable Gate Array (FPGA)

To increase the effective size and measure, the number of directions alternative architecture for SPLD and CPLD have been developed which are called as field programmable gate array (FPGA'S). Field Programmable Gate Array (FPGA) is a flexible architecture programmable logic device. It is a single Very Large Scale Integrated (VLSI) circuit constructed on a single piece of silicon. It consists of identical individually programmable rectangular modulus.

The modulus are separated in both horizontal and vertical directions by horizontal and vertical metallic conductors called channels. In addition, each module has vertical and horizontal conductors at its input and output that cross one or more of the channels.

The content of the modules depends on the type of FPGA, for easy use the modules need to be programmable into the gates and sequential elements. A module may have both combinational and sequential elements and components.

Architecture of FPGA

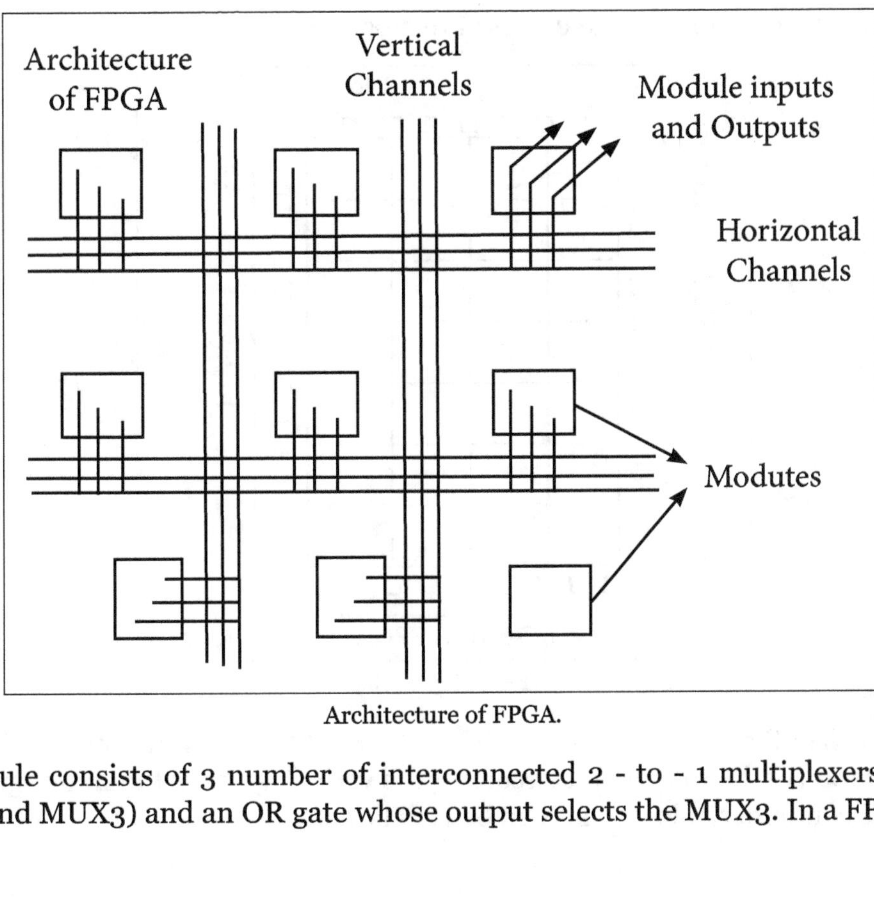

Architecture of FPGA.

The module consists of 3 number of interconnected 2 - to - 1 multiplexers (MUX 1, MUX2, and MUX3) and an OR gate whose output selects the MUX3. In a FPGA, there is no distinct input and output stages as in the AND-OR array and macro cell of a PAL and GAL respectively consider an exemplary FPGA with logic cell with all the logic cells arranged in the array form 12×8 or 24×32 matrix are examples of the array structures. A logic cell is also called as the array structures. A logic cell is also called CLB (Configurable Logic Block). A 12 × 8 possess total 96 logic cells.

A typical logic cells consists of the following:

- The gates to implement SOP function.
- A D-FF with preset and clear.
- Data path selector multiplexer (MUX) at the input.

A typical logic cell is shown in the figure.

- Two number six input AND gates, four number two input AND gates, three number 2 to 1 MUXs and clock input edge triggered DFF with the preset and clear inputs.
- Three number outputs from the ANDs, one output from the MUX and one output from Dff. Per logic cell, the total number of inputs are 21 and outputs are 5. We can not only obtain SOP function on the inputs but also the multiplex and decoding functions at the inputs at the logic cell.

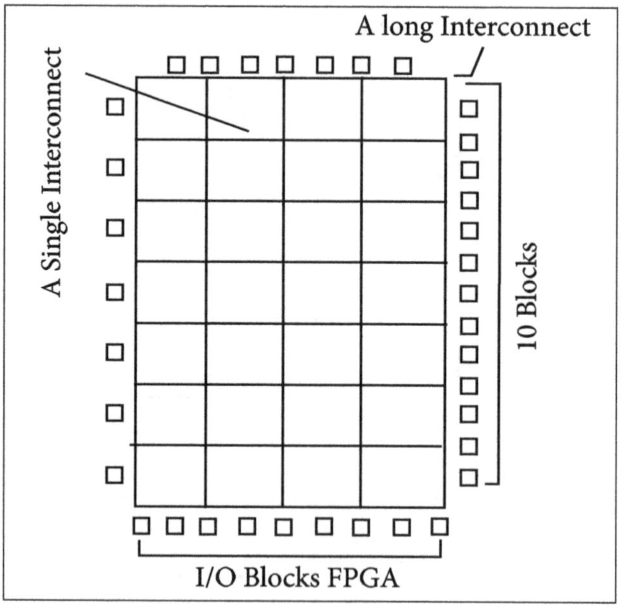

FPGA logic cell.

The DFF in all of the FPGA in configurable as JK FF, RS FF or FFF which facilitates on implementation of the 16 or 32 bit circuit for the various types of the shift registers and counters. The number of logic cells can be very large.

3.4 IC Logic Families

Logic families can be categorized according to the technologies they are built with. In earlier days we had a vast number of technologies which are listed below:

- DL: Diode Logic.
- DTL: Diode Transistor Logic.
- HTL: High threshold Logic.
- RTL: Resistor Transistor Logic.
- TTL: Transistor Logic.
- I^2L: Integrated Injection Logic.
- ECL: Emitter Coupled Logic.
- MOS: Metal Oxide Semiconductor Logic (PMOS and NMOS).
- CMOS: Complementary Metal Oxide Semiconductor Logic.

Since NAND and NOR gates are universal gates, they alone can be used to construct the entire digital circuit instead of different gates.

Factors for Measuring the Performance of Digital Logic Families

- Speed of Operation (Propagation delays).
- Power dissipation.
- Current and Voltage parameters.
- Noise Immunity.
- Operating Temperature.
- Power supply requirements.
- Fan-in.

If the supply voltage is increased on the propagation delay of the CMOS gates, propagation time decreases.

The characteristics of IC digital logic families are usually compared by analyzing the circuit of basic gate in each family. The most important parameters that are evaluated and compared are fan-out, power dissipation, and propagation delay and noise margin.

- Fan-out specifies the number of standard loads that the output of a gate can drive without impairing its normal operation.
- Power dissipation is the supplied power required to operate the gate. This parameter is expressed in milliwatts (mW) and represents the actual power dissipated in the gate.
- Propagation delay: It is the average transition delay time for a signal to propagate from input to output when the binary signals change in value.
- Noise margin is the maximum noise voltage added to the input signal of a digital circuit that does not cause an undesirable change in the circuit output.

S. No	Parameter	CMOS	TTL	ECL
1.	Device used	n-channel & p-channel MOSFET	BJT	BJT
2.	Propagation delay	70 ns	10 ns	500 ps
3.	Power Dissipation	0.1 mW	10 mW	25 mW
4.	Fan-out	50	10	25
5.	Basic gates	NAND-NOR	NAND	NOR - OR
6.	Noise immunity	1.7 V	1 V	0.2V

The basic circuit of the TTL logic family is the NAND gate.

The Schottky TTL is a later improvement that reduces the propagation delay but results in an increase in power dissipation. The fan out of standard TTL is 10. The noise margin is better than 0.4V.

The basic circuit of the ECL family is the NOR gate. The special advantage of ECL gates is their low propagation delay. The power dissipation is high and the noise margin is low ECL offers the highest speed of any family.

The basic circuit of CMOS is the inverter from which both NAND and NOR gates can be constructed. The special advantage of CMOS is extremely low power dissipation. The one major disadvantage of CMOS is its high propagation delay:

- Diode Transistor Logic (DTL) is a class of digital circuits that is the direct ancestor of transistor-transistor logic. It is called so because the logic gating function (e.g. and) is performed by a diode network and the amplifying function is performed by a transistor (in contrast with RTL and TTL).
- Active Power Dissipation: The average power dissipation of a CMOS device whose output is continuously changing is called active power dissipation.

- Completely specified function: All the combinations of states and inputs and their output values are specified.
- Incompletely specified function: Sequential circuits in which some of the states or output values are left.

RTL

A RTL (Register Transfer Level) description describes a circuit registers and the sequence of transfers between these registers but does not describe the hardware used to carry out these operations.

Design Procedure for Register Transfer Language

The basis of RTL design is that circuits can be thought of as a set of register and a set of transfer function defining the data paths between registers.

The first stage of the design has to be specific at a system level that is to be achieved and logic operations on data coming in at the primary inputs of the circuits. At this stage there is no hardware implementations that can be used as the formal specification of the design.

The second stage of the design is to transform the system level design in to an RTL design. It is rare for a design to be directly implemented in exactly the same form as the system level model.

Basic Steps in using RTL

- Identify the data operations.
- Determine the type and precision of the operations.
- Decide what data processing resources to provide.
- Allocate operations to resources.
- Allocate registers for intermediate results.
- Design the controller.
- Design the reset mechanism.

The third stage of the design is to synthesize the RTL design.

The resulting gate-level netlist or schematic can be simulated against the RTC design to confirm that the synthesized circuit has the same behavior.

Finally, the netlist or schematic produced by synthesis is supplied to the placement and routing tools for circuit layout.

RTL in VHDL

VHDL is used to describe the architecture and behavior of discrete electronic system.

Gate levels modeling is lower level of abstraction in which the digital design is described in terms of the logic gates such as AND, OR etc.

Logic synthesis is a process of converting a high-level design into an optimized gate level description by incorporating the given technology library and certain design constraints.

A technology library is provided by the IC fabrication foundry and contains a set of basic logic gates such as AND, OR etc., micro cells such as adders, multiplexers and sequential logic elements such as flip flops.

For the purpose of logic synthesis first the design are written in HDL at register transfer level.

Example:

```
Architecture RTL of NAND is,
begin
process (A, B)
begin
if (A = '1') and (B = '1') then
z<= 0
else
z<= 1
end if
end process
end RTL.
```

The meaning for the RTL statement, T_1: ACC ← ACC and MDR: At time T_1, signal ACC is assigned as an operation of ACC and MDR.

Diode Transistor Logic (DTL)

Diode Transistor Logic or DTL was the first commercially available IC logic family in

53/73 series. A basic DTL circuit consists of diode AND or OR gate circuits followed by a transistor Inverter (NOT- gate) to obtain NAND or NOR logic operations. Figure (a) shows the circuit diagram of a basic DTL behaving like a two-input NAND gate.

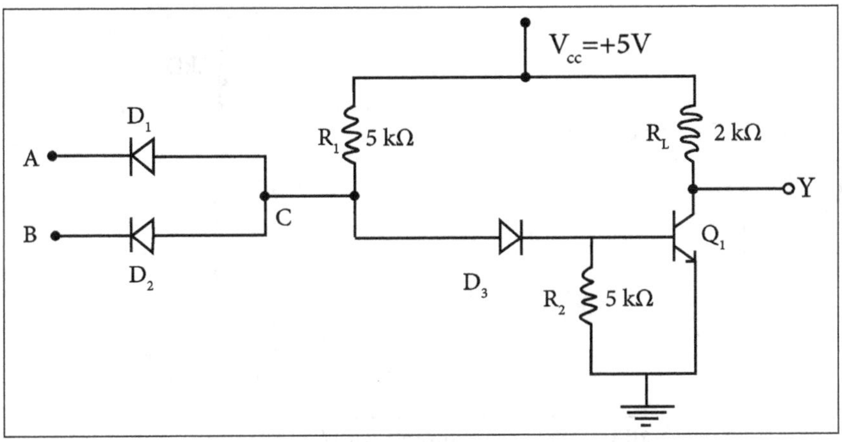

(a) Basic DTL as NAND gate.

The diodes D_1, D_2 and resistor R_1 form an AND gate. If any input A or B is low, the corresponding diode D_1 or D_2 conducts through V_{cc} and R_1. The potential at point C will now be equal to the sum of low input voltage level and forward voltage drop across the diode of 0.7 V. This voltage is insufficient to drive the current through series diode D_3 and base-emitter junction of transistor Q_1. Thus the transistor Q_1 remains cut-off and causes the output voltage at collector point of the circuit to be high, approaching the potential of the supply voltage V_{cc} of +5 V.

If both inputs A and B are high, diodes D_4 and D_2 cannot pass current. The series diode D_3 conducts passing current through it and through the base-emitter junction of transistor Q_1 and holds it into saturation. The input diodes D_1 and D_2 are reverse biased and are off and so the base current is sufficient to drive the transistor into saturation, due to which the output at Y drops to low level.

DU gates can be de-signed with either 21kΩ or 6kΩ load resistor R_1, permitting the designer to vary the power dissipation in a wide range from 15mW to 60mW. But, lower power dissipation using load resistor R_1 of 6kΩ produces greater propagation delays.

The DTL circuit has its various specifications as follows:

- Power Supply voltage Vcc = +5 V.
- Propagation delay = 30 n sec.
- Noise margin = Maximum of 0.7 V.
- Fan-out = 8.

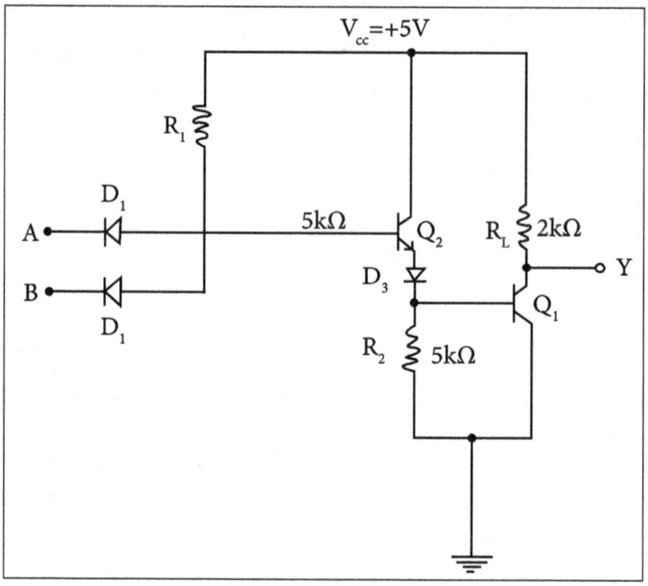

(b) Modified DTL gate, suitable for IC implementation.

The fan-out DTL is limited by the maximum current, which can flow in the collector of transistor Q_1 and can be increased by replacing series diode D_3 in the base circuit by another transistor Q_2 or by just adding another transistor Q_2 before the series diode D, as shown in Figure (b), which is a modified form of DTL gate, suitable for IC implementation. This circuit shown in figure (b) works as a NAND gate. An extra diode D_4 can be connected in series with D_4 to increase the noise margin of the gated and to improve the speed.

Characteristics of TTL Logic

- Power dissipation is usually 10 mW per gate.
- Propagation delays are 10 nS when driving a 15 pF/400 ohm load.
- Voltage levels range from 0 to Vcc where Vcc is typically 4.75V - 5.25V. Voltage range 0V - 0.8V creates logic level 0. Voltage range 2V - Vcc creates logic level 1.

Comparison of CMOS and TTL

- CMOS components are typically more expensive than TTL equivalents.
- CMOS technology is generally less expensive on a system level due to CMOS chips being smaller and needs less regulation.
- CMOS circuits do not draw as much power as TTL circuits while at rest. However, CMOS power consumption increases faster with higher clock speeds than TTL does.

- Due to longer rise and fall times, the transmission of digital signals becomes simpler and less expensive with CMOS chips.
- Lower current draw requires less power supply distribution therefore causing a easy and cheaper design.
- CMOS components are more susceptible to damage from electrostatic discharge than TTL components.

Characteristics of TTL Family

- Low propagation delay.
- Low Power.
- High Speed.
- Noise Immunity.

TTL with Open Collector Output Configuration

The multiple emitters in transistor θ_1 are connected to the inputs, a resistor connected to $+V_{cc}$ must be inserted externally to the IC package for the output to pull up to high voltage level when θ_3 is off otherwise the output acts as an open circuit.

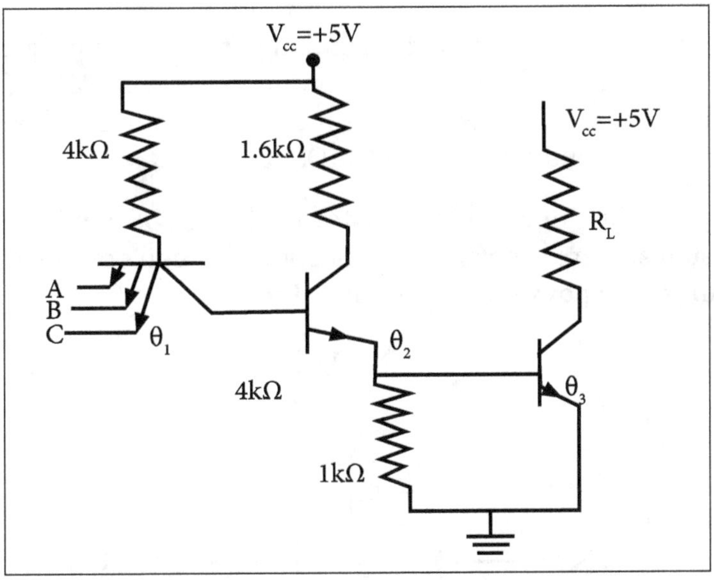

TTL with open collector output configuration.

The two voltage levels of the TTL gate are 0.2v for the low level and from 2.4v to 5v for high level. If any input is low, the corresponding base emitter junction in θ_1 is forward biased. The voltage at the base of θ_1 is equal to the input voltage 0.2v plus a V_{BE} drop of

0.7v or 0.9v. In order to turn ON θ_3, atleast 1.8v is required. So the output transistor θ_3 is OFF and output Y is at $+V_{cc}$.

If all the inputs are high, both θ_2 and θ_3 conduct and saturate. The base voltage of θ_1 is equal to the voltage across its base collector PN junction plus two VBE drops in θ_2 and θ_3. The base emitter junctions of θ_1 are reverse biased and so θ_1 is off.

If the outputs of several open-collector TTL gates are tied together with a single external resistor, a wired AND logic is performed.

The important consideration in using CMOS devices is to:

- Prevent static electric damage.
- Avoid static electricity at all costs.
- Use an earthed soldering iron.
- Solder quickly.
- Power supply voltage in the range 3-15V.
- Don't overload the outputs.

CMOS Logic

Because of the low static power dissipation and high noise immunity, now CMOS logic families is most preferred in large scale integrated circuits. CMOS (Complementary Metal Oxide Semiconductor) has complementary and symmetrical NMOS and PMOS transistors. Figure represented below is a CMOS inverter.

Based on the input value, only one transistor of the CMOS inverter will be ON at a time. So in both states, there is no direct connection between power supply and ground, thereby decreasing static power loss of a transistor.

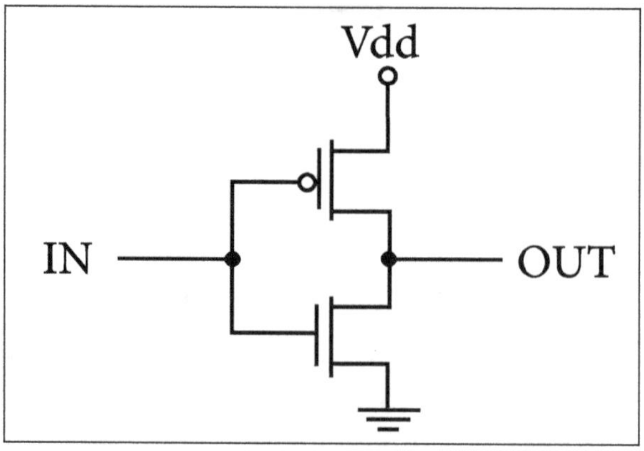

CMOS inverter.

Two Input CMOS NAND Gate

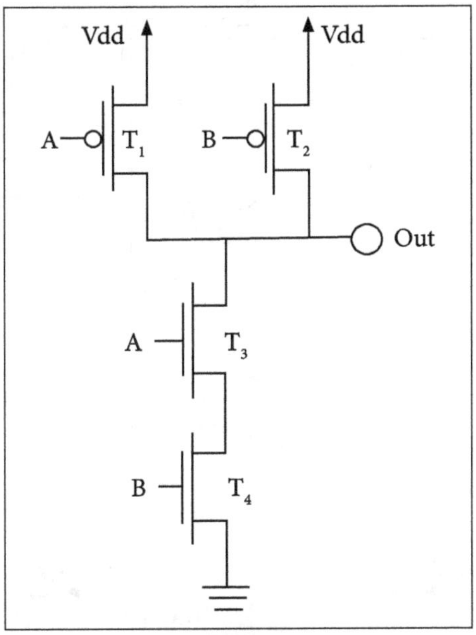

CMOS NAND gate.

When both inputs of a CMOS NAND gate are at high, then transistors T_1, T_2 will be ON and transistors T_3, T_4 will be in OFF state. Now there is a direct connection between output and ground through transistor T_3 and T_4. Output is pulled down to zero.

When any of the input is low, either transistor T_3 or T_4 will be off. This breaks the connection between the ground and the output terminal. At the same time, one of the PMOS transistors is ON, so there is connection between power supply and the output terminal. Now output is pulled up to logic high.

Integrated Injection Logic (I^2L)

Integrated Injection Logic eliminates the need for any resistors, capacitors or transistor isolation. This enables an extremely compact logic circuit to be formed which has low power consumption while maintaining the normal speed of transistor-transistor logic. It is built with multiple collector bipolar junction transistors. Although the logic levels are very close (High: 0.7V, Low: 0.2V), I^2L has high noise immunity, because it operates by current instead of voltage. It is also known as merged-transistor logic.

I^2L Configuration

Consider the DCTL gate structure as shown in the below figure in which there are two logical variables which are assumed to be the outputs of similar DCTL gates and we need to generate the functions $\overline{A+B}$, $\overline{A+\overline{B}}$, $\overline{\overline{A}+B}$, and $\overline{\overline{A}+\overline{B}}$.

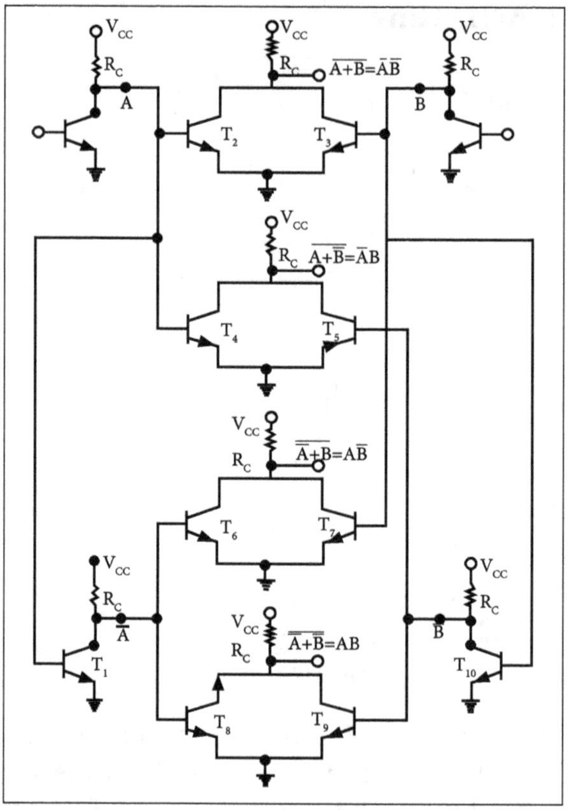

DCTL gate structure for generating functions of two logical variables.

We observe that the bases of transistors T_1, T_2, and T_4 are connected together, and also their emitters are connected together. Therefore, the combination of T_1, T_2, and T_4 can be replaced with a single transistor having one base, one emitter and three collectors. Similarly, other transistors with common-bases are replaced with the multiple-collector transistors. Using this concept, the above figure is redrawn as shown in the below figure.

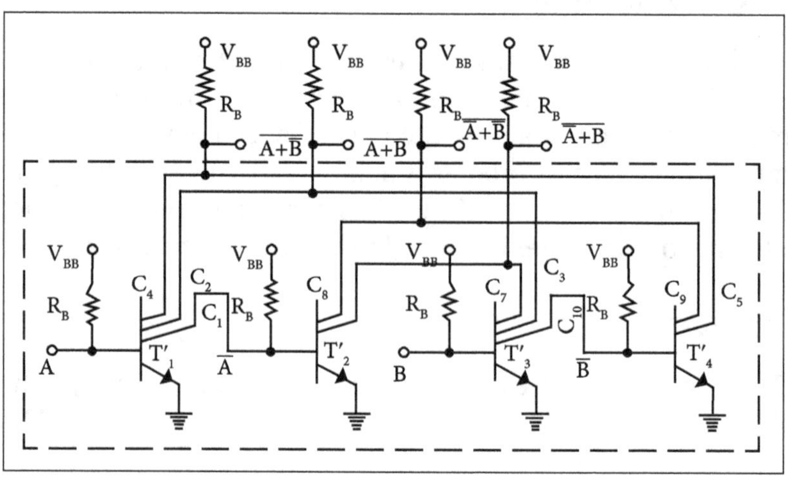

To achieve this, the collector resistors of driving gates are treated as the base resistors of multiple-collector transistors (T_1' and T_3'). Similarly, the collector resistors of T_1 and T_{10} are treated as the base resistors of T_2' and T_4', respectively. Correspondingly, the supply voltages are represented as V_{BB}. The portion of the circuit shown outside the dotted box is either a part of other gates driven by the outputs shown or is omitted altogether.

This means that an I^2L circuit has open-collector outputs, which either feed another I^2L circuit or are to be connected to the supply voltage through the resistors. Suitable values of supply voltage and resistor are to be used for getting proper output voltage levels, for driving other gates such as TTL.

3.4.1 A-D Converters and D-A Converters

A-D and D-A are two types of Data converters, the natural state of audio and video signals is analog. When digital technology was not yet around, they are recorded or played back in analog devices like vinyl discs and cassette tapes. The storage capacity of these devices is very limited and doing multiple runs of re-recording and editing produces the poor signal quality.

Developments in digital technology like the CD, DVD, Blu-ray, flash devices and other memory devices addressed these problems.

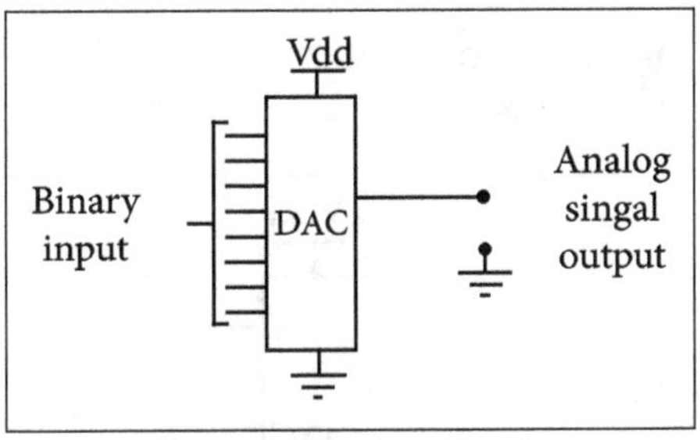

DAC block diagram.

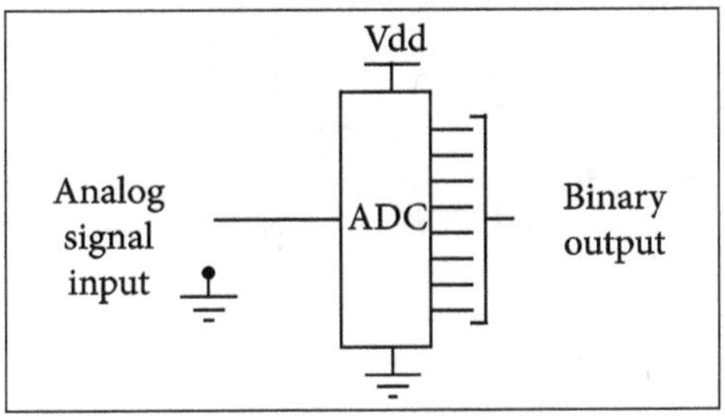

ADC block diagram.

For these devices to be used, the analog signals are first converted to digital signals using analog to digital conversion (ADC). For the recorded audio and video signals to be heard and viewed again, the reverse process of digital to analog conversion (DAC) is used. ADC and DAC are also used in interfacing digital circuits to analog systems.

Typical applications are control and monitoring of temperature, water level, pressure and other real-world data. An ADC inputs an analog signal such as current or voltage and it outputs the digital signal in the form of a binary number. Similarly, DAC inputs the binary number and outputs the corresponding analog voltage or current signal.

Types

The different types of A/D converters are:

- The parallel comparator A/D converter.
- Successive Approximation type.
- Single slope ADC.
- Dual slope A/D converter.

The types of D/A converters are:

- Weighted resistor type.
- R-2R ladder type.

A/D Converters

An ADC inputs an analog signal such as current or voltage and it outputs the digital signal in the form of a binary number.

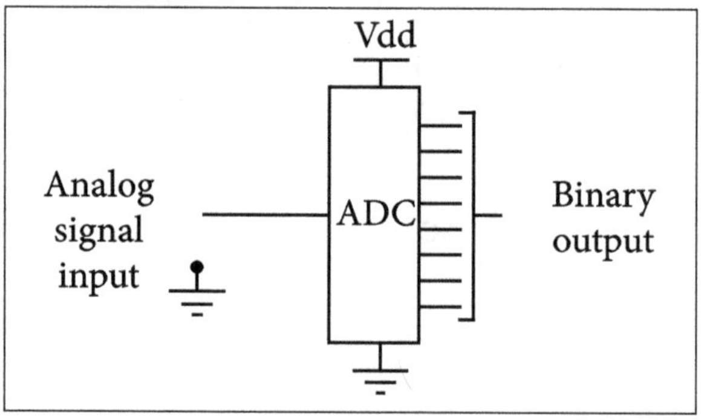

ADC block diagram.

Specifications

Accuracy

Accuracy is the total error with which the A/D converter can convert a known voltage, including the effects of quantization error, gain error, offset error and nonlinearities. Technically, accuracy should be traceable to known standards (for example, NIST) and is generally a "catch-all" term for all static errors.

Resolution

Resolution is the number of bits, N, out of the A/D converter. Resolution determines the size of the least significant bit and thus determines the dynamic range, the code widths and the quantization error.

Dynamic Range

Dynamic range is the ratio of the smallest possible output (the least significant bit or quantum voltage) to the largest possible output (full-scale voltage), mathematically $20 \log_{10} 2^N \approx 6N$.

Offset Error

Offset error is the deviation in the A/D converter's behavior at zero. The first transition voltage should be 1/2 LSB above analog ground. Offset error is the deviation of the actual transition voltage from the ideal 1/2 LSB. Offset error is easily trimmed by calibration.

Gain Error

Gain error is the deviation in the slope of the line through the A/D converter's end points at zero and full scale from the ideal slope of $2N/V_{FS}$ codes-per-volt. Like offset error, gain error is easily corrected by calibration.

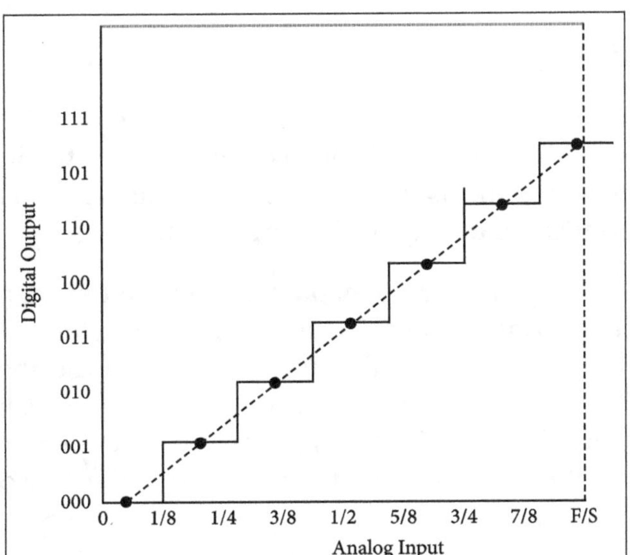

Offset and gain error.

Differential Nonlinearity

Differential nonlinearity (DNL) is the deviation of the code transition widths from the ideal width of 1 LSB. All code widths in the ideal A/D converter are 1 LSB wide, so the

DNL would be zero everywhere. Some data sheets list only the "maximum DNL". The wide range of code widths are shown in below figure.

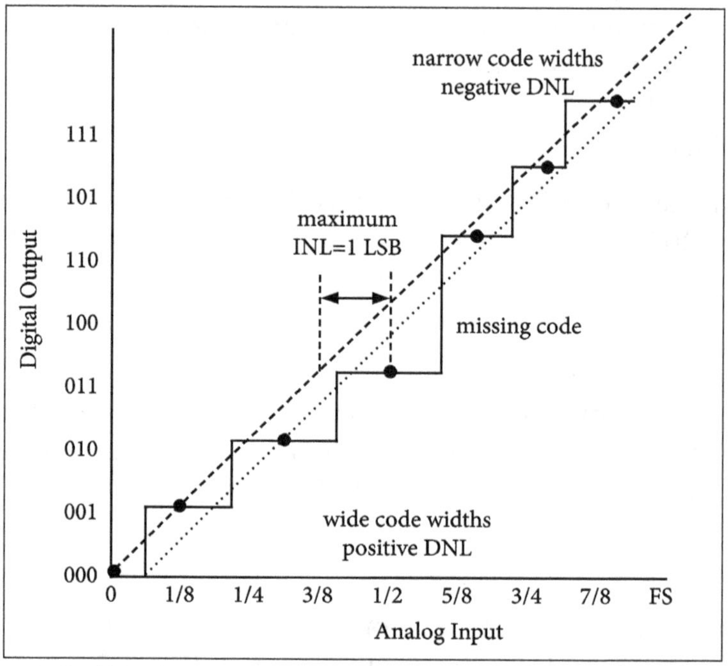

Nonlinearity errors and a missing code. The dashed line is the ideal characteristic and the dotted line is the best fit.

Integral Nonlinearity

Integral nonlinearity (INL) is the distance of the code centers in the A/D converter characteristic from the ideal line. If all code centers land on the ideal line, the INL is zero everywhere. Some data sheets list only the "maximum INL". The deviations of the code centers from the ideal line is shown in the above figure.

Note that there are two possible ways to express maximum INL, depending on the definition of the "ideal line". Data sheet INL numbers can be decreased by quoting maximum INL from a "best fit" line instead of the ideal line. In the above figure, the ideal line (shown dashed) exhibits an INL of 1 LSB, while the best-fit line (shown dotted) exhibits an INL of half that size. While this prevarication underestimates the effect of INL on total accuracy, it is probably a better reflection of the A/D converter's linearity in AC-coupled applications.

Missing Codes

Missing codes are output digital codes that are not produced for any input voltage, usually due to large DNL. In some converters, missing codes can be caused by non-monotonicity of the internal D/A. The large DNL in the above figure causes code 100 to be "crowded out".

Parallel Comparator (Flash) A/D Converter

The Parallel comparator (Flash) A/D converter is the simplest possible A/D converter. Fastest and most expensive technique. Figure shows a 3-bit A/D converter.

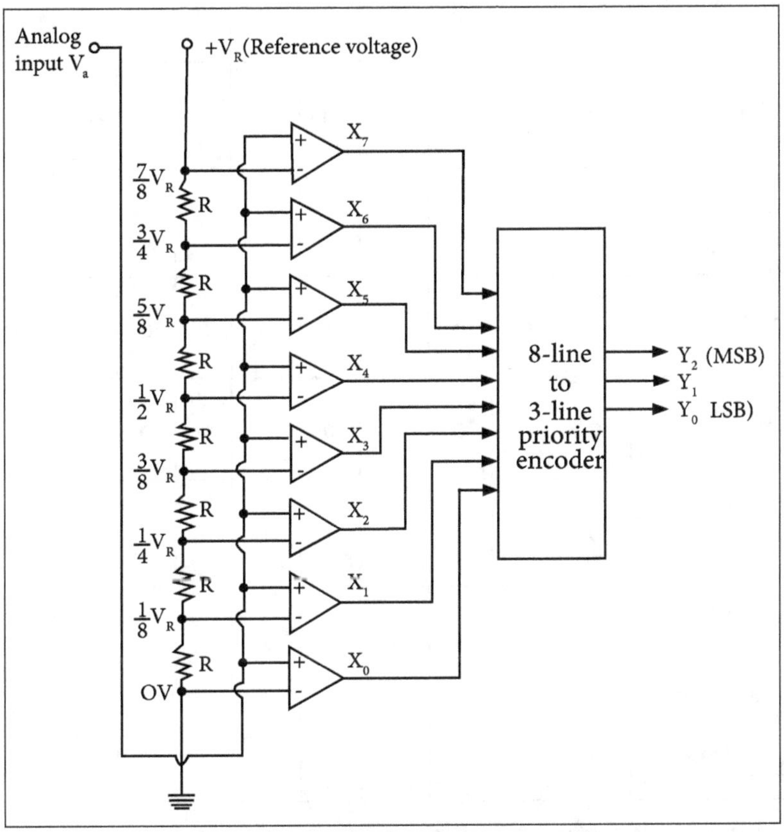

Basic circuit of a flash type A/D converter.

The circuit consists of a resistive divider network, 8 op-amp comparators and a 8 line to 3 line encoders (3-bit priority encoder). The comparator and its truth table is shown below.

Voltage input	Logic output X
$V_a > V_d$	X=1
$V_a > V_d$	X = 0
$V_a = V_d$	Previous Value

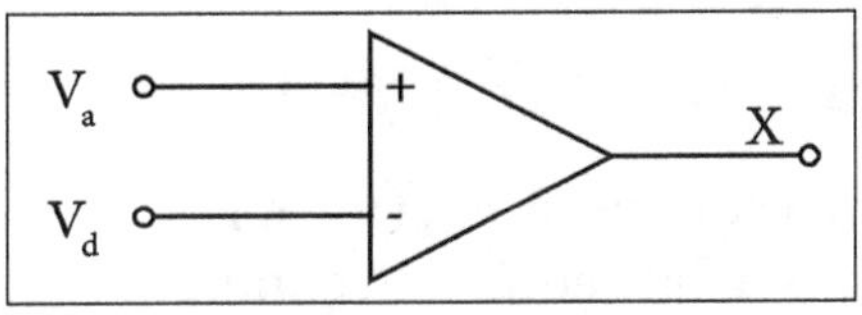

Comparator circuit.

Truth table:

Input voltage V_a	X_7	X_6	X_5	X_4	X_3	X_2	X_1	X_0	Y_2	Y_1	Y_0
$0\text{ to }\frac{V_R}{8}$	0	0	0	0	0	0	0	1	0	0	0
$\frac{V_R}{8}\text{ to }\frac{V_R}{4}$	0	0	0	0	0	0	1	1	0	0	1
$\frac{V_R}{4}\text{ to }\frac{3V_R}{8}$	0	0	0	0	0	1	1	1	0	1	0
$\frac{3V_R}{8}\text{ to }\frac{V_R}{2}$	0	0	0	0	1	0	0	0	0	1	1
$\frac{V_R}{2}\text{ to }\frac{5V_R}{8}$	0	0	0	1	0	0	0	0	1	0	0
$\frac{5V_R}{8}\text{ to }\frac{3V_R}{4}$	0	0	1	1	1	1	1	1	1	0	1
$\frac{3V_R}{4}\text{ to }\frac{7V_R}{8}$	0	1	1	1	1	1	1	1	1	1	0
$\frac{7V_R}{4}\text{ to}V_R$	1	1	1	1	1	1	1	1	1	1	1

At each node of the resistive divider, voltage available at the nodes are equally divided between the reference voltage V_R and the ground. The purpose of the circuit is to compare the analog input voltage V_a with each of the node voltage. The truth table for the flash type A/D converter is shown in Table.

The circuit has the advantages of high speed as the conversion takes place simultaneously rather than sequentially. Typical conversion time is 100 ns or less. Conversion time is limited only by the speed of the comparator and of the priority encoder. By using an advanced micro devices AMD686A comparator and a T1147 priority encoder, conversion delays of the order of 20 ns can be obtained.

Disadvantages

- More no. of comparators required.
- A 2-bit ADC requires 3 comparators, (2n–1) comparator. i.e., a 3-bit ADC needs 7 comparator, whereas 4-bit require 15 comparator. In general, no. of comparators required are 2n-1, where n is the derived number of bits.

- Larger the value of n, more complex is the priority encoder.

SINGLE SLOPE ADC

It consists of a ramp generator and BCD or binary counters. Figure shows the single slope ADC. At the start, reset signal is provided to the ramp generator and the counter. Thus, counters are reset to 0's. The analog output voltage V_{in} is applied to the positive terminal of the comparator. As this is more positive than the negative input, the comparator output goes high.

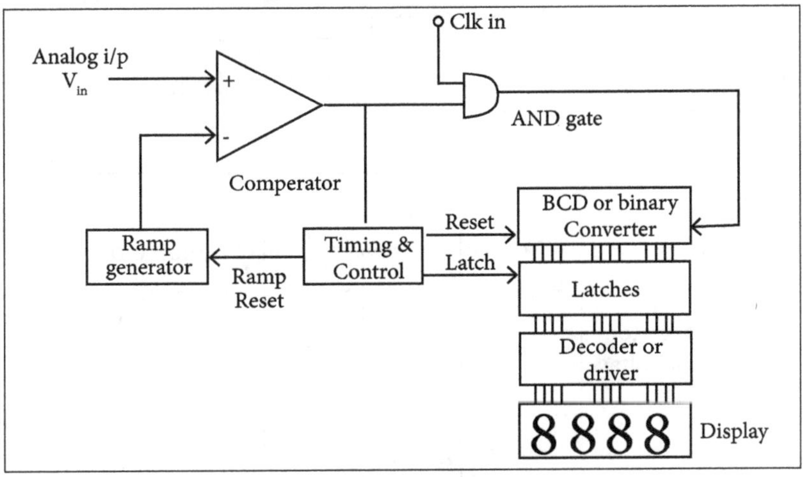

Single slope ADC.

The output of the ramp generator is applied to the negative terminal of the comparator. The high output of the comparator enables the AND gate, which allows circuit to reach to the counter and enabling in turn the high output to start the ramp. The ramp voltages becomes positive until it exceeds the input voltage. When it exceeds, the comparator output goes low. This in turn disables the AND gate which in its turn stops the clock to the counters.

The control circuitry provides the latch signal which is used to latch the counter data. The reset signal resets the counter to 0's and also resets the ramp generator. The latched data is then displayed using decoder and a display device.

DUAL SLOPE A/D CONVERTER

Dual slope conversion is an indirect method for A/D conversion where an analog voltage and reference voltage are converted into time periods by an integrator and then measured by counter.

The above figure shows the functional diagram of the dual slope or dual ramp converter. The analog part of the circuit consists of a high input impedance buffer A_1, precision integrator A_2 and a voltage comparator. The converter first integrates the analog input signal for a fixed duration of 2^n clock periods as shown in the figure. Then, it integrates an internal reference voltage V_R of opposite polarity until the integrator output is zero.

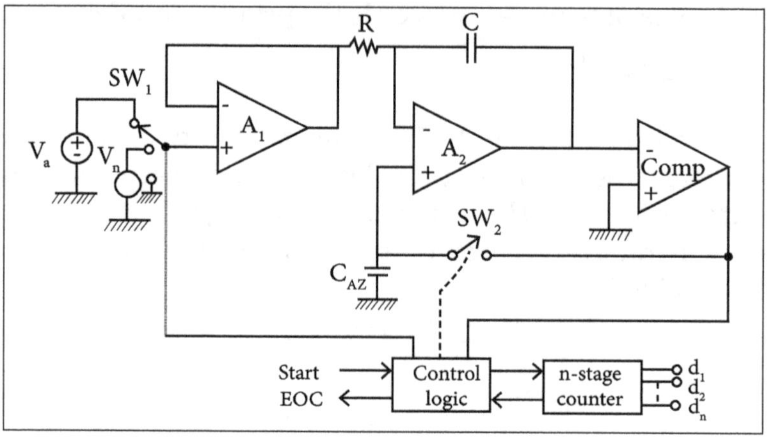

Functional diagram of the dual slope ADC.

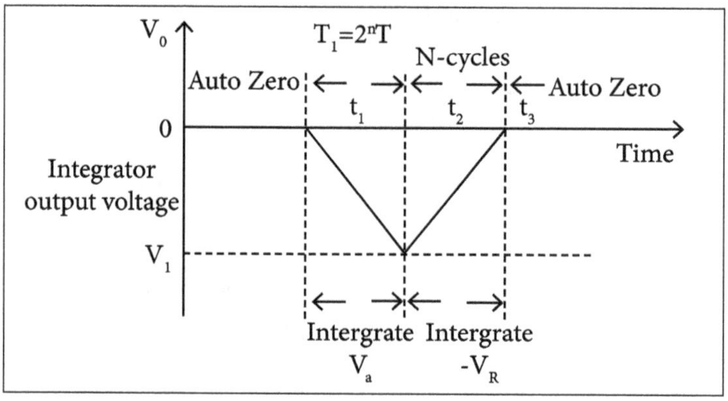

Integrated output waveform for the dual slope ADC.

The number N of clock cycles required to reduce the integrator to zero is proportional to the value of V_a averaged over the integration period. Hence N represents the desired output code. The circuit operates as follows.

Operation

Before the START command arrives, the switch SW_1, is connected to ground and SW_2 is closed. Any offset voltage present in the A_1, A_2, and comparator loop after integration appears across the capacitor till the threshold of the comparator is achieved.

The capacitor C_{AZ} thus provides automatic compensation for the input offset voltages of all the three amplifier. Later, when SW_2 opens, C_{AZ} acts as a memory to hold the voltage required to keep the offset null. At the arrival of the START command at $t = t_1$, the control logic opens SW_2 and connects SW_1 to V_a and enables the counter starting from zero. This circuit uses an n-stage ripple counter and therefore the counter resets to zero after counting 2^n pulses.

The analog voltage V_a is integrated for a fixed number 2^n counts of clock pulses after which the counter resets to zero. If the clock period is T, the integration takes place for a time $T_1 = 2^n \times T$ and the output is a ramp going downwards as shown in the figure. The counter resets itself to zero at the end of the internal T_1 and the switch SW_1 is connected to the reference voltage ($-V_R$). The output V_o will now have a positive slope. As long as V_o is negative, the output of the comparator is positive and the control logic allows the clock pulse to be counted.

However, when V_o becomes just zero at time $t = t_3$, the control logic issues an end of conversion (EOC) command and no further clock pulses enter the counter. It can be shown that the reading of the counter at t_3 is proportional to the analog output voltage V_a.

In the figure,

$$T_1 = t_2 - t_1 \frac{2^n \text{ counts}}{\text{Clock rate}}$$

$$t_3 - t_2 = \frac{\text{digital count N}}{\text{Clock rate}}$$

For an integrator,

$$\Delta V_o = \left(\frac{1}{RC}\right) V(\Delta t)$$

The voltage V_o will be equal to V_1 at the instant t_2 and can be written as,

$$V_1 = \left(\frac{1}{RC}\right) - V_a (t_2 - t_1)$$

The voltage V1 is also given by,

$$V_1 = \left(\frac{1}{R_C}\right) - V_R (t_2 - t_3)$$

$$Va(t_2 - t_1) = V_R (t_3 - t_2)$$

Putting the values of $(t_2 - t_1) = 2^n$ and $(t_3 - t_2) = N$.

We get,

$$V_a (2^n) = V_R (N)$$

$$V_a = (V_R)(N/2^n)$$

Since VR and n are constant, the analog voltage V_a is proportional to the count reading N and is independent of R, C and T.

The dual slope ADC integrates the input signal for a fixed time. Hence, it provides excellent noise rejection of ac signals whose periods are integral multiplies of the integration time T_1. The main disadvantage of the dual slope ADC is the long conversion time. Dual slope converters are particularly suitable for accurate measurement of slowly varying signals such as thermocouples and weighing scales. Dual slope ADC's also form the basis of digital panel meter and multimeter.

D/A Converter

DAC inputs the binary number and outputs the corresponding analog voltage or current signal.

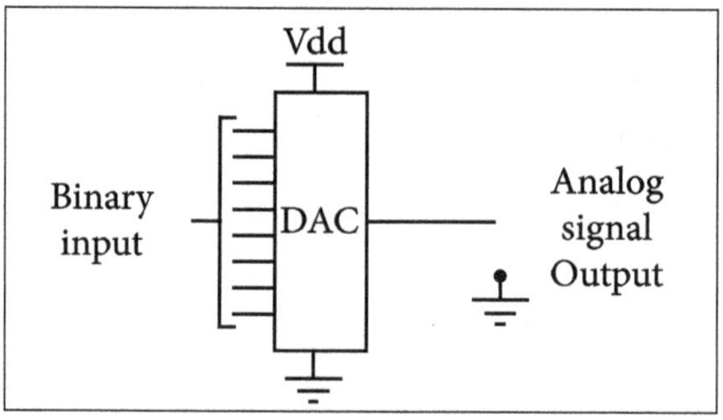

DAC block diagram.

Specifications

Resolution (Output Step Size)

It is given by the output voltage range divided by the number of possible output levels. An N-bit DAC can output 2^N different levels in 2^N-1 steps. N ranges from 6 (for simple high-speed converters) up to 20 or more (for precision instrumentation).

Accuracy

It is the maximum difference between the output and the theoretical output. This is approximately the same as the resolution.

Settling Time

It is the time it takes the output to reach a percentage (e.g. 99%) of the final value. Values can range from hundreds of milliseconds for high-precision units to nanoseconds for devices used in high-frequency waveform generators.

Linearity

It is the maximum difference between the output and a linear interpolation between the minimum and maximum values. This is usually is in the order of $1/2^N$.

Monotonic

The output is monotonic if an increase in the digital input always causes an increase in the analog output. This is usually guaranteed.

There are various types of digital DAC interfaces. Some devices have chip-select and address pins and can be easily interfaced to microcomputers as if they were peripheral chips. Other devices omit the registers and thus require additional logic to interface to a microprocessor.

There are various types of DAC output circuits. Some are designed to generate a current (rather than a voltage) that is proportional to the digital input. Often the DAC is simply a set of N switches and an accurate voltage divider network and an external voltage reference must be supplied.

Finally, there are different ways of manufacturing DAC devices. In addition to the standard monolithic IC's, there are also hybrid devices that mount IC chips together with a small number of discrete devices into an IC-like package. This allows the manufacturer to adjust ("trim") each device during manufacture to obtain higher precision. There are also DAC modules in which a D/A converter built from discrete components is packaged in a sealed module. Hybrids and modules tend to be used for higher-performance devices.

R-2R Ladder Type

Wide range of resistors is required in binary weighted resistor type DAC. This can be avoided by using R–2R ladder type DAC where only two values of resistor are required. It is well suited for integrated circuit realization. The typical value of R ranges from 2.5kΩ to 10 kΩ.

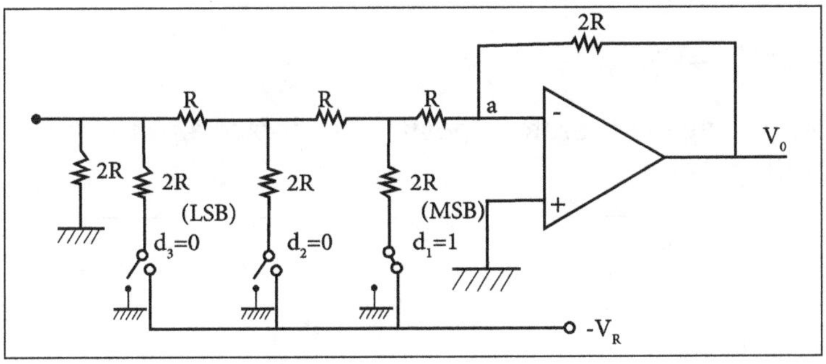

R-2R ladder DAC.

The above figure shows 3-bit DAC where the switch positions d_1, d_2, d_3 corresponds to the binary word 100. The circuit can be simplified to the equivalent form of the figure.

Weighted Resistor Type

One of the simplest circuit shown in Fig. uses a summing amplifier with a binary weighted resistor network. It has n-electric switches d_1, d_2.... d_n controlled by binary input word. These switches are single pole double throw (SPDT) type.

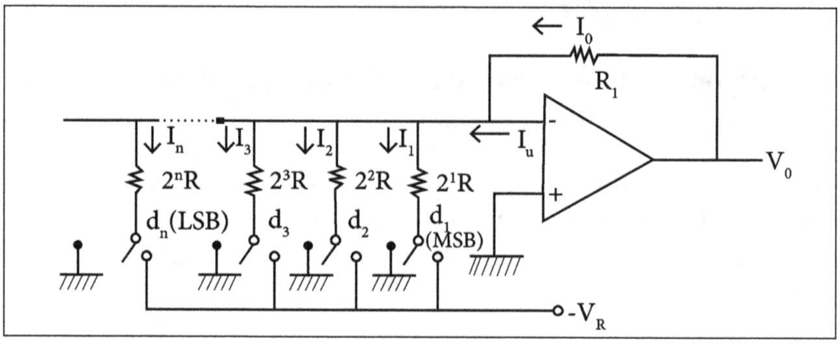

Binary weighted DAC.

If the binary output to a particular switch is 1, it connects the resistance to the reference voltage ($-V_R$). If the input bit is 0, the switch connects the resistor to the ground. In this figure, output current I_o for an ideal op-amp can be written as,

$$I_o = I_1 + I_2 + I_3 .... I_n$$

$$= \frac{V_R}{2R} d_1 + \frac{V_R}{2^2 R} d_2 + ... + \frac{V_R}{2^n R} d_n$$

$$I_o = \frac{V_R}{R}\left[d_1 2^{-1} + d_2 2^{-2} + ... + d_n 2^{-n}\right]$$

The Output voltage, $V_o = I_o R_f$,

$$V_o = \frac{V_R}{R} \times R_f \left(d_1 2^{-1} + d_2 2^{-2} + ... + d_n 2^{-n}\right)$$

Comparing equations, it can be seen that,

If $R_f = R$ and $K = 1$ & $V_{Fs} = V_R$

The analog output voltage is therefore positive staircase as shown in the below figure for a 3-bit weighted resistor.

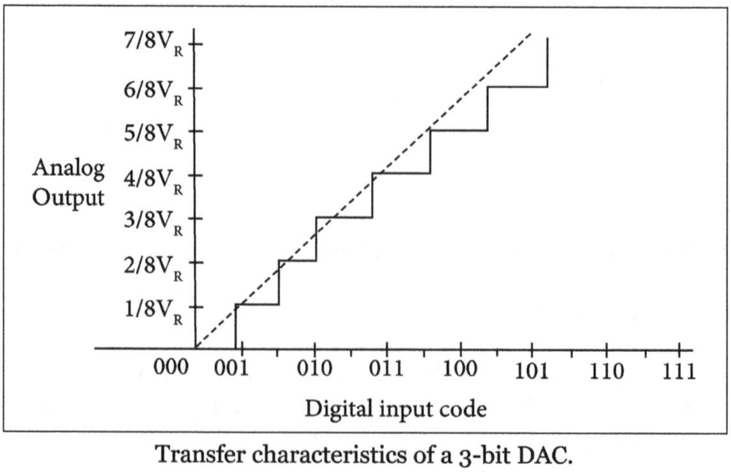

Transfer characteristics of a 3-bit DAC.

- Although the op-amp is connected in inverting mode, it can also be connected in non-inverting mode.
- The op-amp is simply working as a current to voltage converter.
- The polarity of the reference voltage is chosen in accordance with the type of the switch used. For e.g., for TTL compatible switches, the reference voltage should be +5V and output will be negative.

The accuracy and stability of a DAC depends upon the accuracy of the resistors and tracking of each other with temperature.

Disadvantage

Wide range of resistor values are required. For better resolution, the input binary word length has to be increased. Thus, as the number of bit increases, the range of resistance value increases. For a 12-bit DAC, the largest resistance required is 5.12 mΩ. If the smallest is 2.5 kΩ. The fabrication of such a large resistance in IC is not practical.

The voltage drop across a large resistor due to the bias current would also affect the accuracy. For smaller values of resistors, the loading effect may occur. The finite resistance of the switches disturbs the binary-weighted relationship among the various currents particularly in the most significant bit positions, where the current setting resistances are smaller.

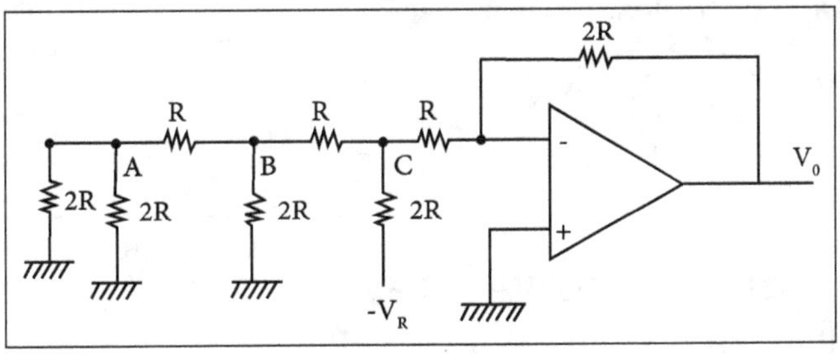

Simplified circuit.

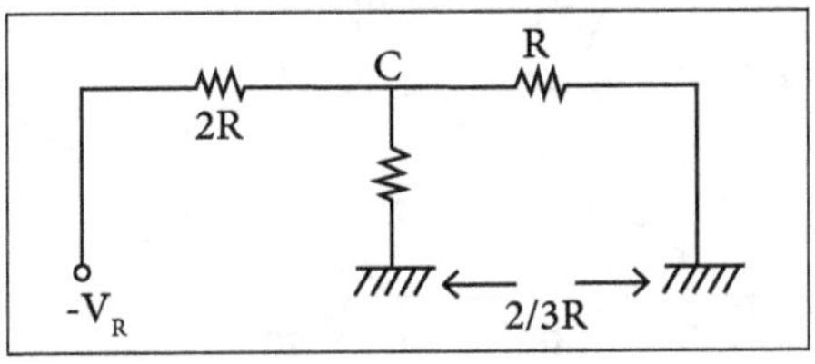

Equivalent circuit.

Then, the voltage at node C can be easily calculated by the set procedure of network analysis.

$$\frac{-V_R\left(\frac{2}{3}R\right)}{2R+\frac{2}{3}R}=\frac{-V_R}{4}$$

The output Voltage,

$$V_o = \frac{-2R}{R}\left(\frac{-V_R}{4}\right) = \frac{V_R}{2} = \frac{V_{FS}}{2}$$

The switch position corresponding to the binary word 001 in 3-bit DAC is shown in the below figure.

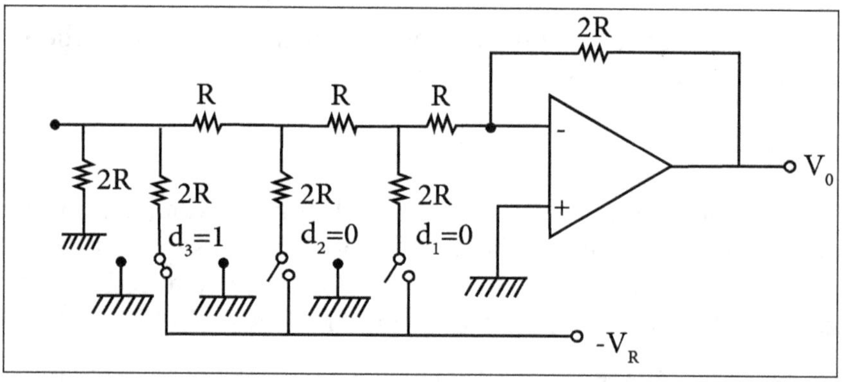

R-2R ladder DAC for switch position 001.

The circuit can be simplified to the equivalent form of the figure. The voltage at the nodes (A, B, C) formed by resistor branches are easily calculated in a similar fashion and the output voltage becomes,

$$V_o = \left(\frac{-2R}{R}\right)\left(\frac{-V_R}{16}\right) = \frac{V_R}{8} = \frac{V_{FS}}{8}$$

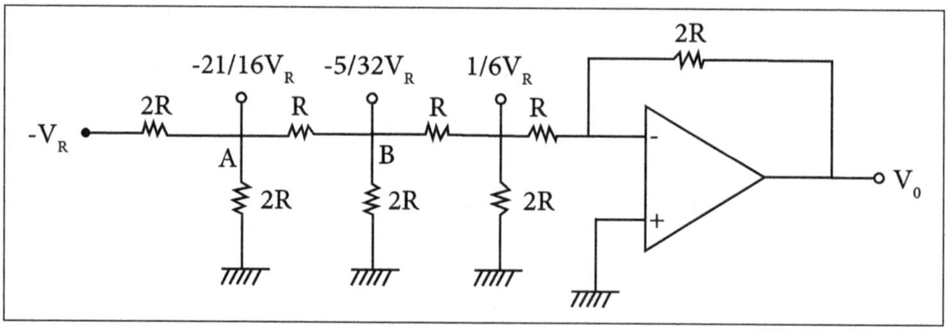

Equivalent circuit.

In a similar fashion, the output voltage for R-2R ladder type DAC corresponding to other 3-bit binary words can be calculated.

Switches for D/A Converters

The inverted ladder DAC's are ideally suited to monolithic fabrication using CMOS technology.

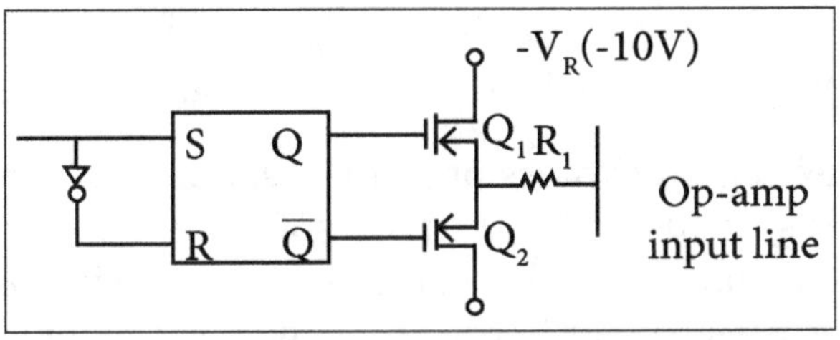

A totem pole MOSFET switch.

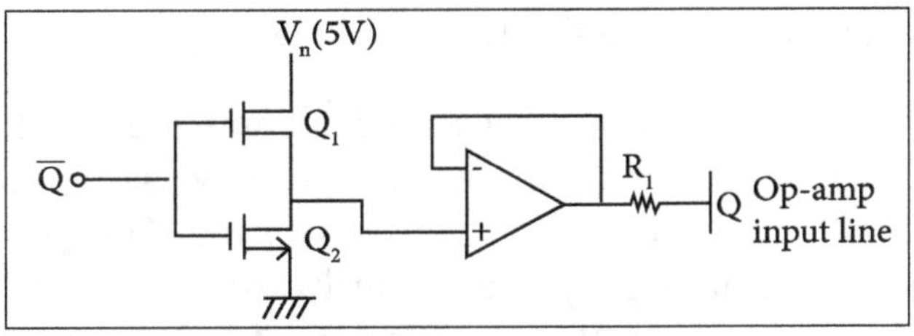

CMOS inverter as switch.

Different types of digitally controlled SPDT electronic switches are available of which two are shown in the above figures.

Two complementary gate inputs Q and $\bar{Q}$ come from MOSFET. SR flip flop or a binary cell of a register which holds one bit of the digital information to be converted to an analog number.

Assume a negative logic, (i.e.) logic '1' in the bit line, S = 1 and R = 0 so that $Q = 1, \bar{Q} = 0$. This drives the transistor Q_1 ON, thus connecting the resistor R to the reference voltage VR whereas the transistor Q_2 remains off, similarly a '0' at the bit line connects the resistor R, to the ground potential Q=1.

Another SPDT switch consists of CMOS converter feeding on op-amp voltage follower which drives R_1 from a very low output resistance.

The circuit is using a positive logic with V(1) = V_R = 5 V and V(0) = 0V. The complement Q of the bit under considerations is applied at the input. Thus Q = 0, makes transistor Q_1 off & Q_2 ON. The output of the CMOS inverter is at logic 1 that is 5V is applied to resistor R, through the voltage follower. And if Q = 1, the output of the CMOS inverter is 0V connecting the resistance R, to ground.

3.5 Basic Hardware Description and Verilog/VHDL Programming Language

Hdl (Hardware Description Language)

In electronics, a hardware description language or HDL is any language from a class of computer languages and programming languages for formal description of digital logic and electronic circuits. It can describe the circuit's operation, its design and organization and tests to verify its operation by means of simulation.

HDLs are standard text-based expressions of the spatial and temporal structure and behaviour of electronic systems. In contrast to a software programming language, HDL syntax and semantics includes explicit notations for expressing time and concurrency, which are the primary attributes of hardware. Languages whose only characteristic is to express circuit connectivity between hierarchies of blocks are properly classified as netlist languages used on electric computer-aided design (CAD).

HDLs are used to write executable specifications of some piece of hardware. A simulation program, which is designed to implement the underlying semantics of the language statements, coupled with simulating the progress of time. This provides the hardware designer with the ability to model a piece of hardware before it is created physically. It is the executability that gives HDLs, the illusion of being programming languages. Simulators capable of supporting discrete-event (digital) and continuous-time (analog) modeling exist and HDLs targeted for each are available.

Design using HDL

The vast majority of modern digital circuit design revolves around an HDL description of the desired circuit, device or subsystem.

HDL and Programming Languages

A HDL is analogous to a software programming language, but with major differences. Programming languages are inherently procedural with limited syntactical and semantic support to handle concurrency. HDLs, on the other hand, can model multiple parallel processes that automatically execute independently of one another.

Any change to the process's input automatically triggers an update in the simulator's process stack. Both programming languages and HDLs are processed by a compiler, but with different goals. For HDLs, 'compiler' refers to synthesis, a process of transforming the HDL code listing into a physically realizable gate netlist.

The netlist output can take any of many forms:

- A "simulation" netlist with gate-delay information.
- A "handoff" netlist for post-synthesis place and route.
- A generic industry-standard EDIF format.

On the other hand, a software compiler converts the source-code listing into a microprocessor-specific object-code, for execution on the target microprocessor. As HDLs and programming languages borrow concepts and features from each other, the boundary between them is becoming less distinct. However, pure HDLs are unsuitable for general purpose software application development, just as general-purpose programming languages are undesirable for modeling hardware.

Yet as electronic systems grow increasingly complex and reconfigurable systems become increasingly mainstream, there is a growing desire in the industry for a single language that can perform some tasks of both hardware design and software programming. System C is an example of embedded system hardware which can be modeled as non-detailed architectural blocks.

The target application is written in C/C++ and natively compiled for the host-development system. The high level of abstraction of System C models is well suited to early architecture exploration, as architectural modifications can be easily evaluated with little concern for signal-level implementation issues.

In an attempt to reduce the complexity of designing in HDLs, which have been compared to the equivalent of assembly languages, there are moves to raise the abstraction level of the design. Companies such as Cadence, Synopsys and Agility Design Solutions are promoting System C as a way to combine high level languages with concurrency models to allow faster design cycles for FPGAs than using traditional HDLs.

Approaches based on standard C or C++ are found in the Catapult C tools from Mentor Graphics and in the Impulse C tools from Impulse Accelerated Technologies. Annapolis Micro Systems, Inc.'s Core Fire Design Suite and National Instruments LabVIEW FPGA provide a graphical dataflow approach to high-level design entry.

Languages such as System Verilog, System VHDL and Handel-C seek to accomplish the same goal, but are aimed at making existing hardware engineers more productive versus making FPGAs more accessible to existing software engineers. Thus, System Verilog is more quickly and widely adopted than System C.

HDL Code for Mealy State Diagram

```
module Mealy-mdl (x, y, CLK, RST);

input x, CLK, RST;

output y;

reg y;

reg [1:0] prstate, Nxtstate;

parameter S0 = 2'b00, S1 = 2'bol, S2 = 2'b10, S3 = 2'b11;

always @ (POS edge CLK or neg edge RST)

if (~ RST) Prstate = S0; // Initialize to state S0
```

```
else Prstate = NxtState; // Clock operations
always @ (Prstate or x); // Determine next state
case (Prstate)
S0: if (x) Nxtstate = S1;
else Nxtstate = S0;
S1: if (x) Nxtstate = S3;
else Nxtstate = S0;
S2: if (~x) Nxtstate = S0;
else Nxtstate = S2;
S3: if (x) Nxtstate = S2;
else Nxtstate = S0;
end case
always @ (Prstate or x) // Evaluate output
case (Prstate)
S0: y = 0;
S1: if(x) y = 1'b0; else y = 1'b1;
S2: if (x) y = 1'b0; else y = 1'b1;
S3: if (x) y = 1'b0; else y = 1'b1;
end case
end module.
```

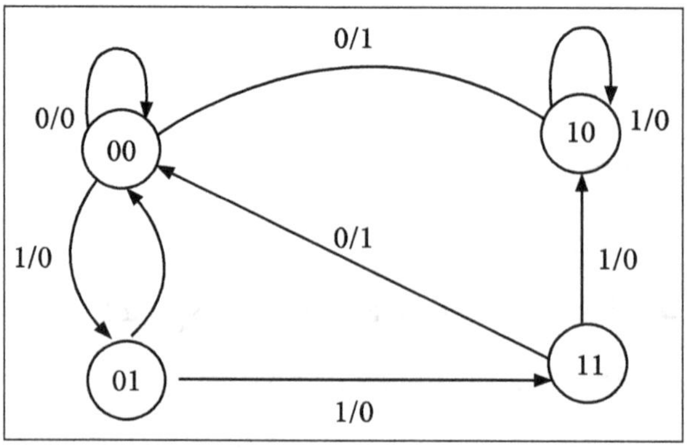

Modeling Techniques in HDL

- Structural Modeling.
- Dataflow Modeling.
- Behaviour Modeling.
- Mixed Modeling.

Types of 'Operators' Supported by VHDL

- Logical Operators.
- Arithmetic Operators.
- Relational Operators.
- Shift Operators.

Arithmetic Operators

*, /, mod, rem, +, –, +, – and (concatenation).

Logical Operators

sll, srl, sla, sra, rol, ror, =, /=, <, <=, >=, > and or nand, nor, xor, xnor.

Use of 'Packages' in VHDL

The 'module' is the basic building block of VHDL. There are some declarations which are common across many design units. A package is the convenient mechanism to store and share such declarations. It is an optional design unit. A set of declarations contained in a package declaration may be shared by many design units. It defines items that can be made visible to other design units.

A package is represented by:

- Package declaration.
- Package body.

Package declaration defines the interface to the package. The items declared in a package declaration can be accessed by other design units by using "Library" and "Use" clauses.

Package body contains the details of a package, that is behavior of the subprograms and the values of the deferred constants are declared in the package declaration. The package body may contain other declarations.

Data Types Supported in VHDL

VHDL supports a variety of data types. The type of variable, signal or constant determines the operators that are predefined for that object as well as the range of values that it can take on. The VHDL data types can be broadly classified in to following five data types.

Scalar Types: The scalar types includes numeric data types and enumerated data types. The numeric type consist of integer, floating point (real) and physical types. Bit, Boolean and Character are all enumerated types.

- Composite Types: Array and record types are composite data types. The values of these types are collection of their elements.
- Access Types: They are pointers. They provide access to objects of a given data type.
- File Types: They provide access to object that contains a sequence of values of a given type.
- Other Types: They include the data types provided by the several external libraries.

3.5.1 VHDL Program of Logic Gates, Adders, Subtractors, Multiplexers, Comparators, Decoders and Shift Registers

A test bench is a VHDL model that verifies the design functioning. It uses files as input. It is used for testing, counting the errors, verification and stimulation. It consists of a set of stimulus and stimulator. The objective of test bench is to generate stimulus to test the system, apply them to the system under test, monitor the output for correctness and log the transactions in a human readable form. System under test can also be termed as unit under test (UUT).

VHDL Test Benches

An HDL description that provides the stimulus to a design is called a test bench. A test bench is an HDL program used for describing and applying a stimulus to an HDL model circuit in order to test it and observe its response during simulation.

Test benches can be quite lengthy and complex thus it may take longer time to develop than the design that is being tested. The results of a test are as good as the test bench that is used for testing a circuit.

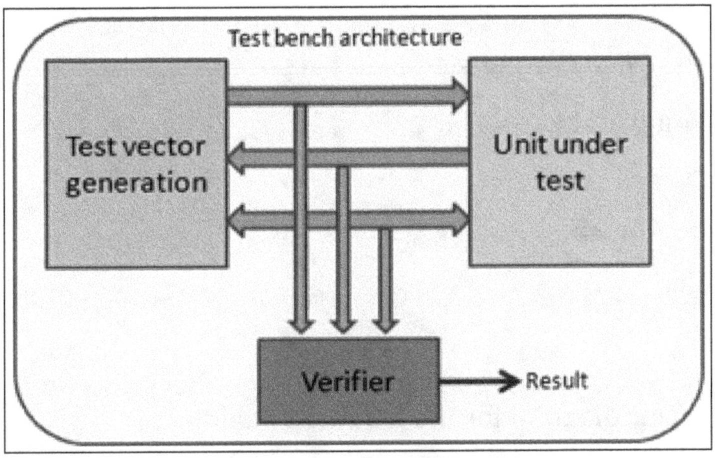

Architecture of test bench.

Test vector generator generates the stimulus for the UUT. This modules can also take outputs coming from the UUT to interactively change the stimulus. Verifier takes the stimulus and the responses from UUT and checks for correctness. Simple test bench will not have verifier modeled into it. So the designer must manually check the correctness of the UUT output.

Test Bench Strategies

As the test bench becomes a part of the hierarchy in our code, the following is recommended:

- Making the test bench the top level of the code.
- The test bench instantiates the unit under test (UUT) and stimulus is applied from the top-level test bench to the lower-level design or portion of the design being tested.
- Using the instance name UUT for the instantiated unit under test.
- This is the default instance name that Project Navigator expects.

Example

```
*/Test bench for simple_circuit_prop_delay/*

module t_simple_circuit_prop_delay;

Wire D, E;

reg A, B, C;

Simple_circuit_prop_delay M1 (A, B, C, D, E);

Initial
```

```
begin
A= 1'bO; B = 1'bO; C = 1'bO;
#100A = l'bl; B = l'bl; C = l'bl;
end
initial #200$ finish;
end module.
```

Programs

1. Let us draw the logic diagram for the following module.

```
module request (A, B,C, Q, CLK);
input A, B, C, CLK;
Output Q;
request Q, E;
always @ (posedge CLK)
begin
E ⇐ A & B
Q ⇐ E/C.
end
end module.
```

Solution:

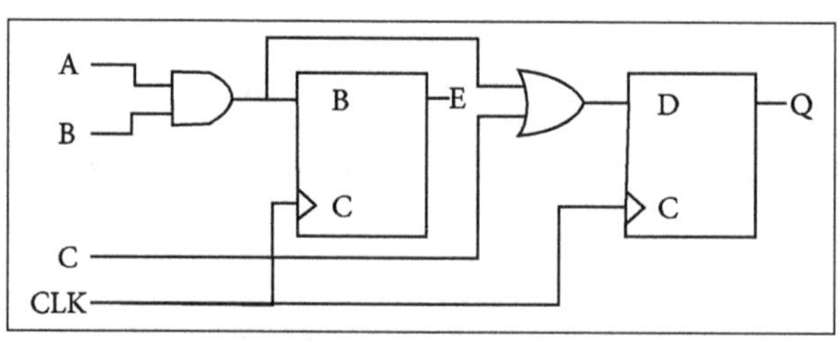

2. Let us see the HDL description for the following circuit.

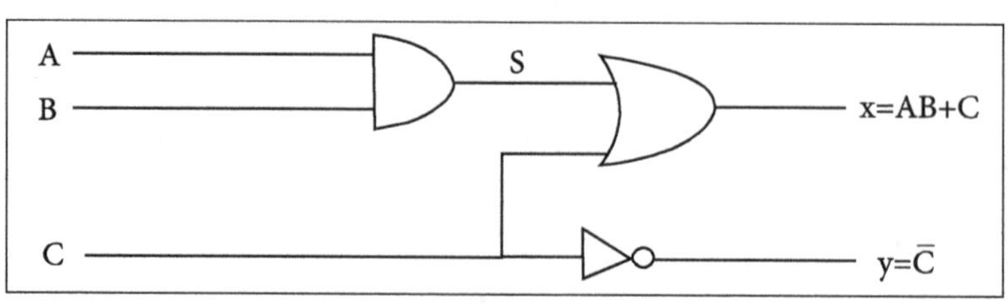

```
Module Circuit (A, B, C, x, y);
Input A, B, C;
output x, y;
wire S;
assign s = A and B;
assign x = SIC;
assign y = ~c;
end module
```

3. VHDL code for AND gate.

```
    LIBRARY IEEE;
    USE IEEE.std_logic_1164.all;
    ENTITY and gate IS
    PORT (A, B: IN STD_LOGIC;
    Y: OUT STD LOGIC);
    END and gate;
ARCHITECTURE arch_and of and gate IS
    BEGIN
    Y <= A and B
    END;
```

4. HDL for half adder

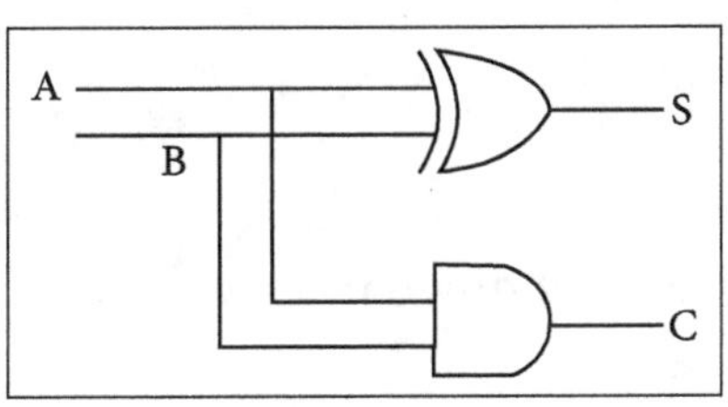

Half adder.

```
Library IEEE;
use IEEE.Std_logic_1164.all;
entity HA is
```

```
Port (A, B: in Std_logic;
S, C: out Std_logic);
end HA;
architecture dataflow of HA is
begin
S = XOR B;
C = A and B;
end dataflow;
```

5. HDL for four bit adder.

Four bit adder:

```
library IEEE;
use IEEE. std_logic_1164.all;
entity adder 4 is
port A, B: in std_logic_vector (3 downto 0);
C1: in std_logic
S: Out std_logic_vector (3 downto 0);
Co: Out std_logic;
end adder A;
architecture structure of adder 4 is
component fulladder
port (x, y, cin: in std_logic);
end component;
signal c: std_ logic_vector (3 down to 1);
begin
fa0 = fulladder Portmap (A(0), B(0), Ci C(1), S(0));
fa1 = fulladder Portmap (A(1), B(1), C(1), C(2), S(1));
fa2 = fulladder Portmap (A(2), B(2), C(2), C(3), S(2));
```

fa3 = fulladder Portmap (A(3), B(3), C(3), Co, S(3));

end structure.

6. HDL Program for full adder.

LIBRARY IEEE;

USE IEEE.Std_logic_1164.all;

ENTITY full add IS

PORT (C_{in}, A, B; IN STD_LOGIC;

sum, C_{out}: OUT STD_LOGIC);

END full add;

ARCHITECTURE logic Func OF fulladd IS BEGIN

SUM ⇐ A XOR B XOR C_{in};

Cout ⇐ (A AND B) OR (C_{jn} AND A) OR (C_{jn} AND B);

END Logic Func;

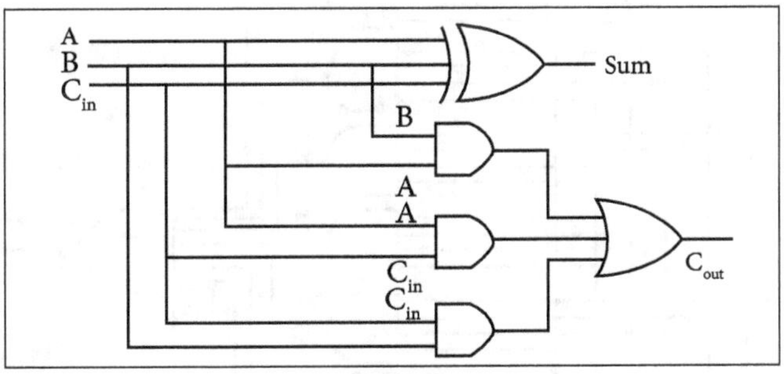

Full adder.

7. HDL code that implements an 8: 1 multiplexer.

module multiplexer (D, S_0, S_1, S_2, Y);

input [0: 7] D, S_0, S_1, S_2;

output y;

Wire S_0 not' S_1not' e, f, g, h, i, j, k, l;

not

n_1 (S_0 not' S_0);

n_2 (S_1 not' S_1);

$N_3(S_2$ not' $S_2)$;

and

a_1 (e, D [0], S_2 not' S_1 not' S_0 not);

a_2 (f, D [1], S_2 not' S_1 not' S_0 not');

a_3 (g, D [2], S_2 not' S_1 not' S_0 not');

a_4 (h, D [3], S_2 not' S_1 not' S_0);

a_5 (k, D [4], S_2 not' S_1 not' S_0 not');

a_6 (j, D [5], S_2 not' S_1 not' S_0);

a_7 (k, D [6], S_2 not' S_1' S_0 not');

a_8 (l, D [7], S_2 not' S_1' S_0);

or (y, e, f, g, h, i, j, k, l);

endmodule.

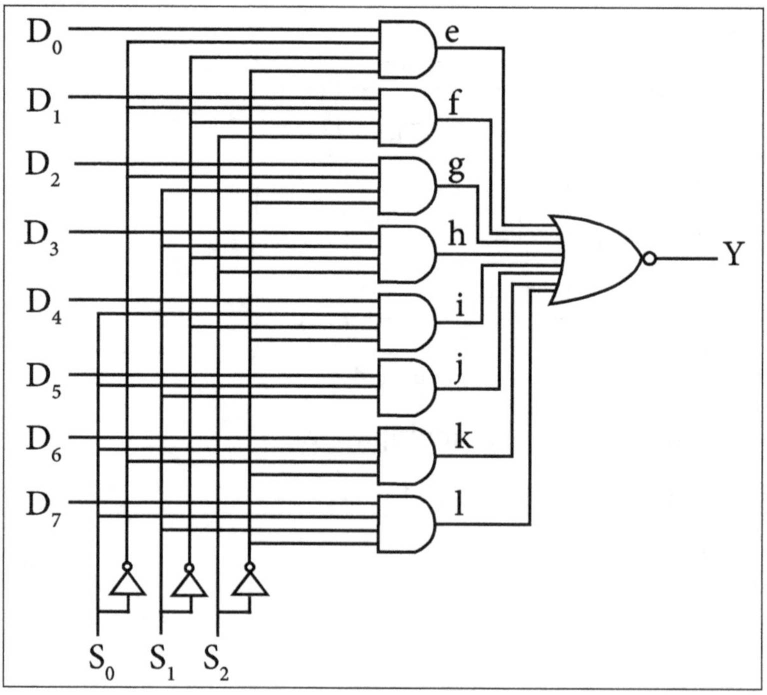

8. VHDL code for 8:1 MULTIPLEXER.

library IEEE;

Use

```
IEEE.std_logic_1164.0.11;
entity mux is
Port (D7, D6, D5, D4, D3, D2, D1, D0; in std-logic;
S: in std_logic_vector (2 down 0);
F: Out std_logic);
end mux;
architecture dataflow of mux is,
begin
F< = (not S (2) and not S (1) and not S(0) and D0) or
(not S(2) and not S(1) and S(0) and D1) or
(not S(2) and not S(0) and S(1) and D2) or
(not S(2) and S(1) and S(0) and D3) or
(not S(1) and not S(0) and S(2) and D4 or
(not S(1) and S(0) and not S(2) and D5) or
(not S(0) and S(1) and not S(2) and D6) or
(not S(2) and S(1) and not S(0) and D7);
end;
```

9. VHDL module listing for a 16:1 MUX.

```
Library IEEE;
Use IEEE.Std_logic_1164.all;
Library UNISIM;
Use UNISIM.V COMPONENTS. ALL;
entity MUX_16_1_SUB m is Port
{
DATA_I: in Std_logic_Vector (15 down to O);
SELECT_i: in Std_logic_Vector (3 down to O);
DATA O: out Std logic
```

```
};
end MUX_16_1_SUBm;
architecture MUX_16_1_SUBm_arch of MUX_16_1_SUBm_ component
declarations:
component MUX F7
Port
{
Io: in Std_logic;
I1: in Std_logic;
S: in Std_logic;
O: out Std_logic;
};
end component;
__ Signal declarations;
Signal DATA MSB:Std logic;
Signal DATA LSB: Std logic;
begin
--
__ if synthesis tools support MUX F7;
__ select_PROCESS: process (select_I, DATA_I)
__ begin
__ case SELECT_I is
__ When '0000' ⇒ DATA_o* DATA_I (0);
__ When '0001' ⇒ DATA_o* DATA_I (1);
__ When '0010' ⇒ DATA_o* DATA_I (2);
__ When '0011' ⇒ DATA_o* DATA_I (3);
__ When '0100' ⇒ DATA_o* DATA_I (4);
```

```
_ _ When '0101' ⇒ DATA_o*DATA_I (5);
_ _ When '0110' ⇒ DATA_o* DATA_I (6);
_ _ When '0111' ⇒ DATA_o* DATA_I (7);
_ _ When '1000' ⇒ DATA_o* DATA_I (8);
_ _ When '1001' ⇒ DATA_o* DATA_I (9);
_ _ When '1010' ⇒ DATA_o* DATA_I (10);
_ _ When '1011' ⇒ DATA_o* DATA_I (11);
_ _ When '1100' ⇒ DATA_o* DATA_I (12);
_ _ When '1101' ⇒ DATA_o* DATA_I (13);
_ _ When '1110' ⇒ DATA_o* DATA_I (14);
_ _ When '1111' ⇒ DATA_o* DATA_I (15);
_ _ When others ⇒ DATA_o* 'X';
_ _ end case;
_ _ end Process SELECT PROCESS;
_ _ if synthesis tools DONOT Support MUX F7;
SELECT_PROCESS_LSB: PROCESS (SELECT_I, DATA_I)
begin
case SELECT_I (2 down to 0) is
When '000' ⇒ _LSB* DATA_I (0);
When '001' ⇒ _LSB* DATA_I (1);
When '010' ⇒ _LSB* DATA_I (2);
When '011' ⇒ _LSB* DATA_I (3);
When' 100' ⇒ _LSB* DATA_I (4);
When '110' ⇒ _LSB* DATA_I (6);
When '111' ⇒ _LSB* DATA_I (7);
When others ⇒ DATA_MSB* 'X';
end case;
```

```
end process SELECT PROCESS LSB;
SELECT_PROCESS_MSB: PROCESS (SELECT_I, DATA_I) begin
case SELECT I (2 down to o) IS
When 'ooo' ⇒ _DATA_MSB* DATA_I (8);
When 'oo1' ⇒ _MSB* DATA_I (9);
When 'o1o' ⇒ _MSB*.DATA_I (1o);
When 'o11' ⇒ _MSB* DATA_I (11);
When '1oo' ⇒ _MSB* DATA_I(12);
When '1o1' ⇒ _MSB* DATA_I(13);
When '11o' ⇒ _MSB* DATA_I(14);
When '111' ⇒ _MSB* DATA_I(15);
When others ⇒ DATA_MSB* 'X';
end case;
end process SELECT_PROCESS_MSB;
_ _ MUX F7 instantiation
U_MUX F7: MUX F7
Port map
{
Io ⇒ DATA_LSB,
I1 ⇒ DATA_MSB,
S ⇒ SELECT I(3)
O ⇒ DATA_O,
};
end mux_16_l_SUBM_arch;
```

10. 4-Bit comparator.

```
LIBRARYIEEE;
USE IEEE.Std_logic_1164.all;
```

```
USE IEEE.Std_logic_unsigned. all;
ENTITY compare IS
PORT (A, B: IN STD_LOGIC_VECTOR (1 DOWN TO Aeq B, Agt B;
A it B = OUT STD_LOGIC);
END compare;
ARCHITECTURE Behavior of compare IS BEGIN
Aeq B ⇐ '1' WHEN A = B ELSE '0';
Agt B ⇐ '1' WHEN A > B ELSE '0';
A lt B ⇐ '1' WHEN A < B ELSE '0';
END Behavior;
```

11. VHDL program of 8 bit comparator.

```
library ieee;
use ieee . std _ logic _ 1164 . all;
entity comparator is
port (A: in std _ logic _ vector (7 downto 0);
B: in std _ logic _ vector (7 downto 0);
less: out std _ logic;
equal: out std _ logic;
greater: out std _ logic;
end comparator
architecture behavioural of comparator is
begin
process (A, B)
begin
if (A < B) then
less < = 'i';
equal < = '0';
```

```
greater <='0';
else if (A = B) then
less <= '0';
equal < = '1';
greater < = '0';
else
less < = '0';
equal < = '0';
greater < = "1";
end if;
end process;
end behavioural;
```

12. Structural VHDL description for a 2 to 4 decoder.

```
library ieee;
use ieee . std_logic_1164. all;
entity decoder is,
Port (X, Y: in std_logic;
D: out std_logic_vector (3 downto 0);
end decoder;
architecture structure of decoder is
component not 1
Port (a: in std_logic; b: out std_logic);
end component;
component and2
Port (X, Y: in std_logic; Z: out std_logic);
end component;
signal X bar, Y bar;
```

```
begin
g1: not 1 portmap (X, X bar);
g2: not 1 portmap (Y, Y bar);
g3: and 2 portmap (X bar, Y bar, D (0));
g4: and 2 portmap (X bar, Y, D (1));
g5: and 2 portmap (X, Y bar, D (2));
g6: and 2 portmap (X, Y, D (3));
end structure;
```

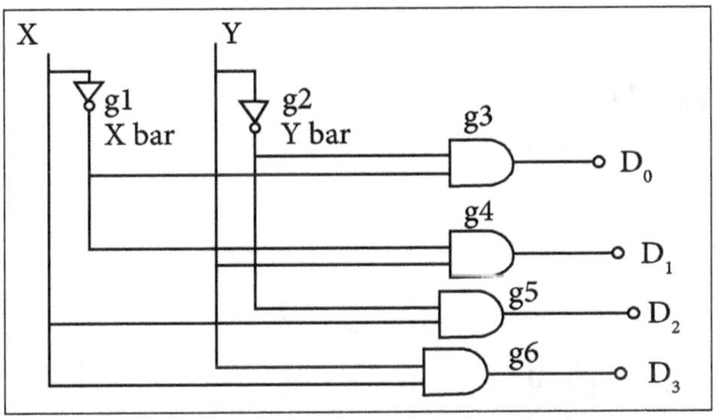

13. VHDL module for a JK flip flop.

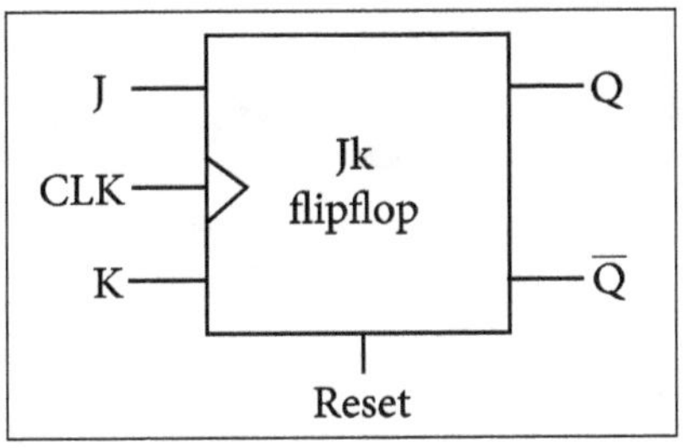

Truth Table:

J	K	CLK	Qn - 1
0	0	↑	Q^n
0	1	↑	0
1	0	↑	1
1	1	↑	Q_n
X	X	↓	Q_n

```
library ieee;
use ieee. std_logic_1164. all;
entity JK_FF is
port (J,K, CLK, Reset: in Std. ulogic;
Q, qbar: out Std ulogic);
end JK_FF;
architecture sig of JK_FF is
Signal state: Std_ulogic;
begin
process (clk, Reset) is
begin
if (Reset = '0') then
State ⇐ '0';
else if rising edge (clk) then
care Std_ulogic_vecto (J,K) is
when "11" ⇒
state ⇐ not state;
when "10" ⇒
state ⇐ '1';
when "01" ⇒
state ⇐ '0';
when others ⇒
null;
end care;
end if;
end process;
Q ⇐ state;
```

```
Q has ⇐ not state;
end sig;
```

14. VHDL code for Decode counter.

```
library ieee;
use ieee.std_logic_1164.all;
entity decounter is
port (CLK: in std_logic;
Q: inout std_logic_vector (3 downto 01: = '0000');
end decounter,
architecture behavioral of decounter is
begin
Process (CLK)
begin
if CLK = '1' and CLK event then
if Q = "1010" then
Q < = "0000";
Else
Q < = Q+ 1;
end if;
end if;
end process;
end behavioral;
```

15. HDL for four bit binary counter with parallel load.

4 bit binary counter:

```
library IEEE;
use IEEE.std_logic_1164.all
entity counter is
```

```
port (CLK, Reset: in std_logic;
Q = inout std_logic_vector (3 downto 0));
end counter
architecture behavioral of counter is
begin
process (CLK, RESET)
begin
if RESET ='1' then
Q ⇐ "0000"
Else if CLK = '1' and CLK' even then
Q ⇐ Q+ 1;
end if
end process
end behavioral
```

16. Full Subtractor Design using Logical Gates (VHDL Code).

```
library IEEE;
use IEEE.STD_LOGIC_1164.all;
entity full_subtractor is
port(
a: in STD_LOGIC;
b: in STD_LOGIC;
c: in STD_LOGIC;
difference: out STD_LOGIC;
borrow: out STD_LOGIC
};
end full_subtractor;
architecture full_subtractor_arc of full_subtractor is
```

```
begin
difference <= a xor b xor c;
borrow <= ((not a) and b) or
(b and c) or
(c and (not a));
end full_subtractor_arc;
```

17. VHDL code for Four Bit Adder-Subtractor.

```
library ieee;
use ieee.std_logic_1164.all;
entity adderSubtractor is
port( mode : in std_logic;
A3, A2, A1, A0: in std_logic;
B3, B2, B1, B0: in std_logic;
S3, S2, S1, S0: out std_logic;
Cout, V: out std_logic);
end adderSubtractor;
--Structural architecture
architecture struct of adderSubtractor is
component xorGate is --XOR component
port ( A, B: in std_logic;
F: out std_logic);
end component;
component Full_Adder is --FULL ADDER component
port( X, Y, Cin: in std_logic;
sum, Cout: out std_logic);
end component;
--interconnecting wires
```

```
signal C1, C2, C3, C4: std_logic; --intermediate carries
signal xor0, xor1, xor2, xor3: std_logic; --xor outputs
begin
GX0: xorGate port map (mode, B0, xor0);
GX1: xorGate port map (mode, B1, xor1);
GX2: xorGate port map (mode, B2, xor2);
GX3: xorGate port map (mode, B3, xor3);
FA0: Full_Adder port map (A0, xor0, mode, S0, C1);-- S0
FA1: Full_Adder port map (A1, xor1, C1, S1, C2); -- S1
FA2: Full_Adder port map (A2, xor2, C2, S2, C3); -- S2
FA3: Full_Adder port map (A3, xor3, C3, S3, C4); -- S3
Vout: xorGate port map (C3, C4, V); -- V
Cout <= C4; -- Cout
end struct;
```

Permissions

All chapters in this book are published with permission under the Creative Commons Attribution Share Alike License or equivalent. Every chapter published in this book has been scrutinized by our experts. Their significance has been extensively debated. The topics covered herein carry significant information for a comprehensive understanding. They may even be implemented as practical applications or may be referred to as a beginning point for further studies.

We would like to thank the editorial team for lending their expertise to make the book truly unique. They have played a crucial role in the development of this book. Without their invaluable contributions this book wouldn't have been possible. They have made vital efforts to compile up to date information on the varied aspects of this subject to make this book a valuable addition to the collection of many professionals and students.

This book was conceptualized with the vision of imparting up-to-date and integrated information in this field. To ensure the same, a matchless editorial board was set up. Every individual on the board went through rigorous rounds of assessment to prove their worth. After which they invested a large part of their time researching and compiling the most relevant data for our readers.

The editorial board has been involved in producing this book since its inception. They have spent rigorous hours researching and exploring the diverse topics which have resulted in the successful publishing of this book. They have passed on their knowledge of decades through this book. To expedite this challenging task, the publisher supported the team at every step. A small team of assistant editors was also appointed to further simplify the editing procedure and attain best results for the readers.

Apart from the editorial board, the designing team has also invested a significant amount of their time in understanding the subject and creating the most relevant covers. They scrutinized every image to scout for the most suitable representation of the subject and create an appropriate cover for the book.

The publishing team has been an ardent support to the editorial, designing and production team. Their endless efforts to recruit the best for this project, has resulted in the accomplishment of this book. They are a veteran in the field of academics and their pool of knowledge is as vast as their experience in printing. Their expertise and guidance has proved useful at every step. Their uncompromising quality standards have made this book an exceptional effort. Their encouragement from time to time has been an inspiration for everyone.

The publisher and the editorial board hope that this book will prove to be a valuable piece of knowledge for students, practitioners and scholars across the globe.

Index

www.ingramcontent.com/pod-product-compliance
Lightning Source LLC
LaVergne TN
LVHW080601210926
848485LV00003B/91